P9-EMP-257

Fluorescence and Phosphorescence of Proteins and Nucleic Acids

Fluorescence and Phosphorescence of Proteins and Nucleic Acids

Sergei V. Konev

Laboratory of Biophysics and Isotopes
Academy of Sciences of the Belorussian SSR
Minsk, USSR

Translated from Russian

Translation Editor
Sidney Udenfriend
Laboratory of Clinical Biochemistry
National Heart Institute
Bethesda, Maryland

PLENUM PRESS · NEW YORK · 1967

Sergei Vasil'evich Konev, senior scientific worker and group leader in the Biophysics and Isotopes Laboratory of the Academy of Sciences of the Belorussian SSR, in Minsk, and the author of 76 technical papers, was born in 1931. After completing his undergraduate work in biology and soil science, in 1954, at the Moscow State University, he did postgraduate research in the University's Biophysics Department. From 1957 to 1959 he worked at the Biophysics Laboratory of the All-Union Research Institute of Animal Husbandry in Moscow, where he and I. I. Kozunin devised a rapid luminescence method for determining the protein content of milk, for which achievement he was awarded the Silver Medal of the Exhibition of Achievements of the National Economy. Since 1959, Konev and his colleagues at the Biophysics and Isotopes Laboratory in Kiev have studied problems of luminescence and primary photophysical processes in protein and cells, and the migration and conservation of electronic excitation energy in protein systems. He has recently been working on mitogenetic radiation of cells and the role of the electron excited states of proteins in dark reactions of vital activity.

Library of Congress Catalog Card Number 67-10309

The original Russian text, published by Nauka i Tekhnika, Minsk, in 1965, has been updated by the author for the American edition.

СЕРГЕЙ ВАСИЛЬЕВИЧ КОНЕВ

ЭЛЕКТРОННО-ВОЗБУЖДЕННЫЕ СОСТОЯНИЯ БИОПОЛИМЕРОВ

ELEKTRONNO-VOZBUZHDENNYE SOSTOYANIYA BIOPOLIMEROV

Printed in the United States of America

FOREWORD

Fluorescence and phosphorescence are proving to be extremely sensitive probes for elucidating conformation of proteins and nucleic acids and for studying molecular interactions. Newer instrumentation and techniques hold forth great promise for the future of these luminescence methods in biopolymer research. It must be noted, however, that the discovery that certain amino acids, purines, and pyrimidines emit fluorescence or phosphorescence is relatively recent, occurring within the last decade. Professor Konev is one of the pioneers in the application of these procedures to biopolymers and is highly qualified to write about this subject.

This book, though written largely as a monograph of the author's own contributions, is also an excellent review of the subject. Of particular interest are the references to many important Russian papers in this field which have not been recognized in the Western literature. It is apparent from this book that fluorescence and phosphorescence methods are being used about as widely in Russia as elsewhere in the world and that we must not overlook these important contributions. Konev's studies on protein fluorescence have been widely recognized. It is of interest to learn about these and other of his applications. The last part of the book, which deals with fluorescence as a means to probe into the structure and conformation of macromolecules in intact cells, is most interesting. Aside from published symposia this book is the first written specifically about luminescence of biopolymers.

Sidney Udenfriend

Bethesda, Maryland
May, 1967

CONTENTS

INTRODUCTION

About eight years have now elapsed since the publication of Reid's "Excited States in Chemistry and Biology," a first-rate book in the depth and breadth of its treatment of the problem. During this time there has been a rapid accumulation of data on the nature and role of electronic excited states in biology. These data were primarily centered around the electronic excited states of pigment systems, the normal biological functioning of which entails interaction of the molecules with light quanta.

In addition, within the last decade a foundation has been laid for the rapid development of research on electronic excited states of biopolymers, primarily proteins and nucleic acids. These investigations have been conducted on a broad front and have included absorption, luminescence, and photochemical methods, and also action-spectrum methods. Although the biological functions of biopolymers in pure form, in contrast to the various types of pigment molecules, do not include interaction with quanta of radiant energy and despite the fact that their everyday functioning is an example of an extreme dark reaction, the pursuit of research in this direction is of definite interest for biology from at least three points of view.

Firstly, biopolymers are subjected, naturally or artificially, to exposure to ultraviolet and ionizing radiation. The topics which can be mentioned in this connection include the whole gamut of radiobiological questions: the effects of artificial and natural sources of penetrating radiation on the organism; the effects of the ultraviolet component of sunlight on microbial, plant, and animal organisms; the use of artificial ultraviolet sources for the control of harmful organisms; the induction of new and valuable genetic characteristics and the elimination of undesirable attributes in microorganisms in mutagenesis and radioselection; and the use of artificial ultraviolet irradiation of farm animals for therapeutic purposes or to increase productivity. In all these cases the chain of events leading to alteration of the biological characteristics or activity of cells and organisms is largely initiated by electronic excited states of proteins and nucleic acids, which are the main acceptors of ultraviolet and penetrating radiation. In fact, any biological process can be transformed (inhibited, stimulated, or altered) through photochemical methods by inducing electronic excited states of proteins and nucleic acids. In view of this it is obvious that a knowledge of the nature of electronic excited states, photophysical processes of energy transfer, and primary photochemical reactions of biopolymers is of supreme importance for the understanding and intelligent control of all these processes.

Secondly, by investigating electronic excited states with the aid of photoabsorption, photoluminescence, and chemiluminescence techniques, we can obtain a great deal of information not only about the number of biopolymer molecules in the system, but also about their qualitative state. As will be shown later, in several cases luminescence studies even provide information about the structural organization (conformation) of macromolecules. When one considers that information of this kind can be obtained almost instantaneously from the interior of the undamaged, normally functioning cell, one realizes the special importance of spectral-luminescence research techniques in biology.

Thirdly, it can be inferred that the characteristics of the dark and light behavior of molecules conceal a deep unity. By investigating details of the electronic excitation of molecules we can learn about the spatial architecture of their electronic levels and the allowed and forbidden states of the clouds of external electrons of the molecules. Moreover, the outermost electrons predetermine the biochemical and biophysical behavior of molecules in darkness. The role of optical transitions in dark activity, the creation of electronic excited states during dark metabolism, and bioluminescence and mitogenetic rays as a means of physical communication between cells are all questions which touch on one of the most exciting and least resolved problems of molecular biology.

HISTORY OF RESEARCH ON ELECTRONIC EXCITED STATES OF PROTEINS

A new physical property of proteins—their ability to luminesce—was discovered in the last decade and has now been fairly thoroughly investigated.

Despite its comparative newness the question of the luminescence of proteins has a fairly long prehistory. Becchari in the eighteenth century was probably the first to encounter the luminescence of protein-rich tissues when he observed the glowing of his own hand in a dark room after a short exposure to sunlight. Stokes observed the fluorescence of horns and nails (cited by Dhéré, 1933, 1935). Regnault in 1858 and Sechenov in 1859 observed the fluorescence of the crystalline lens and skin, and in 1883 Soret described the violet fluorescence of myosin in 1% hydrochloric acid. Helmholtz (1896) observed the fluorescence of vertebrate eyes, and Hess (1911), that of insect and crustacean eyes. Stübel (1911) gave a description of the luminescence of skin.

In the thirties of this century, which were marked by a craze for black light (sources of artificial ultraviolet light), there were many qualitative observations of the luminescence of the most diverse proteinaceous materials. This luminescence was the subject of investigation by such well-known scientists as Svedberg (1925), Liesegang (1926), Pringsheim (1928), Tiselius (1930), and Teichler (1931).

These authors observed the violet, blue, or yellowish-blue color of the fluorescence, its presence in some proteins (albumin, casein, vitellin, elastin, etc.) and absence in others (milk albumin, fibrin), the dependence of its intensity on the state of aggregation (powder or solution), and its enhancement by heat (Pringsheim and Gerngross, 1928) or exposure to ultraviolet rays (Wels, 1928,

1931; Becker and Szendrö, 1931, and Wiegand, 1929). This clearly indicates that the majority of investigators were not dealing with the fluorescence characteristic of proteins as such, but with the fluorescence of impurities or chemically altered components of proteins.

For instance, Wels (1928), using a nephelometer, visually observed blue fluorescence of serum albumin excited by the 365-mμ mercury line. This luminescence cannot be ascribed to the protein itself, since the red end of the absorption band of proteins lies in the shorter-wave region of the spectrum. In addition, this author observed a similar luminescence in amino acids known to be incapable of luminescence, such as glycocoll, glutamic acid, leucine, and alanine. The complete depolarization of the fluorescence and the enhancement by exposure to ultraviolet and X rays also rule out the possibility of the luminescence belonging to proteins. Probably Wels observed nothing more than the universal blue-green fluorescence.

In a number of cases, however, the authors were dealing with the long-wave "tail" of the true ultraviolet fluorescence of proteins, as in the case of the blue fluorescence of boiled eggs (Van Waegeningh and Heesterman, 1932), and the fluorescence of milk (Jenness and Coulter, 1948).

Several investigations in the thirties confirmed what Becchari had observed, that proteinaceous materials could emit an afterglow. In 1933 Hoshijima found that human cartilage, nails, and tendons exhibited a distinct afterglow after irradiation with a quartz lamp, whereas other tissues and organs (liver, heart, etc.) did not produce such an afterglow. Two years later, in 1935, Leighton observed, with the dark-adapted eye, phosphorescence of the skin of the frog's belly and the human hand, with a lifetime of 2 to 4 sec. In 1937 a vast amount of material of biological origin was investigated by Giese and Leighton. They found that the crystalline lens, stomach, kidney, muscles, blood, and dorsal skin of the frog could not produce an afterglow. Sponge chitin gave a short afterglow. On the other hand, such dense, protein-rich materials as fingernails, birds' beaks, bones, and mollusk shells gave a very long afterglow, which lasted for 20 to 25 sec. The authors qualitatively evaluated the excitation spectrum of this emission and found that it was excited most effectively by light of wavelength 280 (!) mμ.

Unfortunately, the authors did not attempt to link the observed emission with any molecular substrate. A definite step forward in the discovery of the ability of proteins to luminesce was made in 1940 by Reeder and Nelson, who investigated many proteins: casein, glutelin, gliadin, blood fibrin, gelatin, ovalbumin, zein, and hair and wool keratin. Their observed bluish-white fluorescence of solid proteins and blue-green fluorescence of dissolved proteins could not be ascribed to the primary luminescence of protein, since the luminescence was effectively excited by the spectral region of 310 to 400 mμ, which lies outside the absorption band of proteins, and became weaker after dialysis of the preparations. The authors were nevertheless the first to observe a causal relationship between the luminescence of proteins and their tryptophan content. In this investigation, for instance, they observed a marked increase in the intensity of the visible fluorescence of proteins on hydrolysis only in the case of tryptophan-containing proteins. In addition, tyrosine and tryptophan were the only amino acids which gave a blue fluorescence after being boiled in dilute hydrochloric acid. Thus, though they correctly linked the secondary, induced fluorescence with tryptophan, Reeder and Nelson did not discover the primary ultraviolet fluorescence of native proteins.

In 1952 Debye and Edwards investigated several proteins and 18 amino acids at liquid-nitrogen temperature and failed to discover visible fluorescence in any of these substances, although they found distinct phosphorescence of aromatic amino acids, proteins, and some microbial cells. Weber in 1953 fully realized the possible nature of the fluorescence of proteins when he postulated, on the basis of general considerations of the properties of aromatic amino acids and the positions of their absorption maxima, that, firstly, the fluorescence of proteins must lie in the ultraviolet region of the spectrum and, secondly, must owe its origin to aromatic amino acids.

In 1955 Szent-Györgyi observed blue fluorescence of the eye lens excited by short-wave ultraviolet light and showed that the carrier was the protein euglobulin. Assuming correctly that the ability to fluoresce is a general property of proteins, as it has been observed in corneal epithelium and skin epithelium, and pointing out that a considerable amount of the fluorescence would extend into the ultraviolet region and hence would not be detected, Szent-Györgyi nevertheless advanced the ill-founded hypothesis that the protein molecule as a whole is the luminescence carrier. A consideration of his experiments in retrospect enables us now to realize that Szent-Györgyi was probably dealing with the long-wave part of the tryptophan fluorescence spectrum of proteins. However, on discovering a considerable reduction in the fluorescence intensity after thermal denaturation, he erroneously attributed this to the existence of energy levels which were common to the whole native protein molecule and disappeared when the conformation of the macromolecule was destroyed. It is now clear that the reduction of fluorescence intensity in Szent-Györgyi's experiments could have been due to the high sensitivity of tryptophan, the main luminescence center, to changes in the protein structure. Thus, in the middle of the fifties the question of the ultraviolet fluorescence of proteins was literally hanging in the air. Hence, it is not surprising that the ultraviolet fluorescence of proteins was discovered independently at this time by three groups of authors in the United States and the Union of Soviet Socialist Republics. In the account of the history of the discovery of the ultraviolet fluorescence of proteins in his well-known monograph, Udenfriend (1962) traced the origin of investigations of the ultraviolet fluorescence of proteins to the studies of Shore and Pardee (1956) and Konev (1957). Yet this is not quite correct historically. Although Vladimirov and I, like Shore and Pardee, did not know at that time of one another's investigations or of the studies of Duggan and Udenfriend (1956), the first measurements of the ultraviolet fluorescence of proteins rightly belong to these last two authors. They obtained fluorescence spectra of human blood serum with a maximum at 360 mμ (with no correction for the spectral sensitivity of the apparatus). They put forward some arguments in support of the view that the luminescence center of serum is tryptophan. The arguments included the similarity of the fluorescence spectra and the similar quenching of tryptophan and blood serum hydrolyzate by such substances as inorganic nitrite, thiosulfate, and hydrogen peroxide.

Soon after this, Shore and Pardee (1956) recorded the fluorescence of proteins and aromatic amino acids, the maxima of which, judged by the filters employed, were somewhere between 300 and 400 mμ. They measured the quantum yields of the fluorescence and obtained underestimates: 1 to 2% for proteins. The reported reduction in the quantum yields of the fluorescence of proteins and aromatic amino acids with reduction in the wavelength of the exciting light must be regarded as an artifact.

During 1956 and 1957 Vladimirov and Konev (Konev, 1956, 1957 and Vladimirov and Konev, 1957, 1959) conducted their first investigations of the ultraviolet luminescence of proteins. According to these measurements, the fluorescence spectra had a distinct tryptophan maximum at about 350 mμ. The excitation spectra showed an aromatic tryptophan maximum at 280 mμ and a second maximum at 240 mμ (Konev, 1957). This maximum was ascribed to peptide absorption, but further investigations showed that it was an artifact. The exclusively aromatic nature of protein fluorescence was conclusively shown by experiments in which no fluorescence could be detected in proteins which contain no aromatic amino acids (clupein, sturin). The observed constancy of the fluorescence spectra of proteins on excitation by different wavelengths showed that, irrespective of the nature of the centers which initially absorbed the light quanta, the same tryptophan emission centers were implicated in the fluorescence (Konev, 1957).

Later the ultraviolet luminescence of proteins became a subject of investigation in many laboratories in the Soviet Union (Vladimirov, Barenboim, Sapezhinskii, Konev et al.) and abroad (Weber and Teale, Szent-Györgyi et al., Cowgill, Steiner, Edelhoch, and many others).

It can now be regarded as established that the formation of fluorescence and phosphorescence spectra of proteins involves only three aromatic amino acids which are capable of luminescence in the free state. These are tryptophan, tyrosine, and phenylalanine. Hence, we shall start the discussion of the luminescence of proteins with the luminescence of these amino acids in the free state.

Chapter 1

ELECTRONIC EXCITED STATES OF MONOMERS

TRYPTOPHAN

Fluorescence Spectra

The fluorescence spectrum of tryptophan in aqueous solution (Fig. 1) is a broad structureless band with a maximum at 348 mμ and a half width of 60 mμ (Teale and Weber, 1957). The shape of the fluorescence spectrum of tryptophan and the position of the maximum are determined mainly by the indole ring of the molecule without any appreciable contribution from the substituent groups, since indole and its various derivatives—indolepropionic acid, tryptamine, serotonin—have practically the same fluorescence spectra.

The weak influence of the substituent groups on the position of the fluorescence band is indicated as well by the small displacement of the fluorescence maximum when the pH of the medium is altered (Vladimirov and Li Chin-kuo, 1962): $R-\underset{NH_3^+}{\underset{|}{C}}-COOH$ has a maximum at 347 mμ, $R-\underset{NH_3^+}{\underset{|}{C}}-COO^-$ has a maximum at 353 mμ, and $R-\underset{NH_2}{\underset{|}{C}}-COO^-$ has a maximum at 360 mμ. Ionization of the imino group (in the excited state) has a much greater effect. In tryptophan

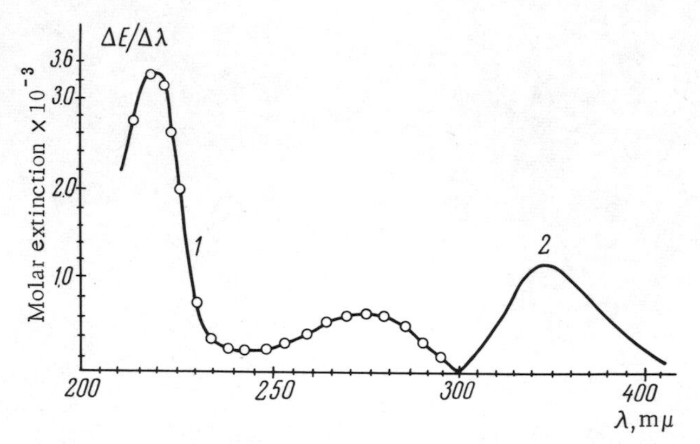

Fig. 1. Excitation spectrum (1) and fluorescence spectrum (2) of tryptophan in aqueous solution (Teale and Weber, 1957).

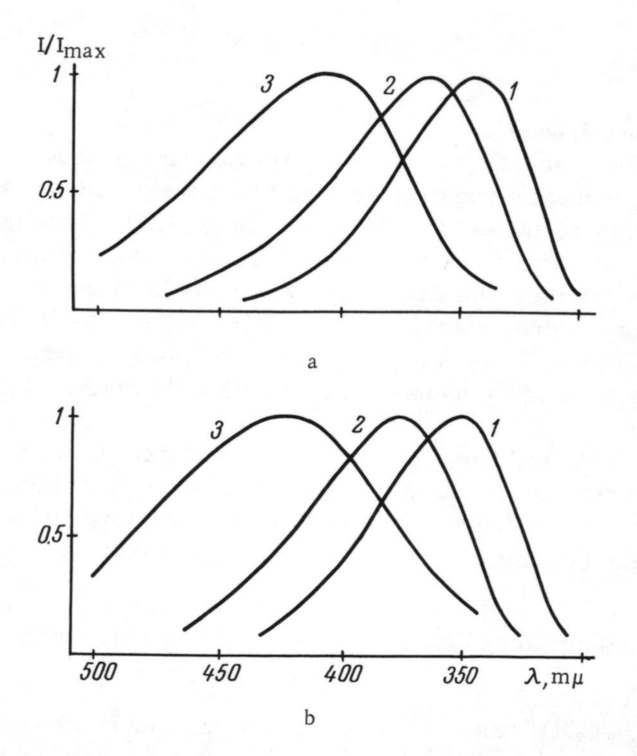

Fig. 2. Fluorescence spectra of (a) indole and (b) tryptophan at room temperature: (1) in neutral aqueous solutions; (2) in mixture of 4% NaOH and 0.5% formaldehyde; (3) in 4% NaOH solution (Chernitskii and Konev, 1965).

in the ionic form (indolyl, $N^{\ominus}$)$-CH_2-\underset{NH_2}{CH}-COO^-$ the fluorescence maximum (Fig. 2) is shifted into the visible region of the spectrum, to 420 mμ (Chernitskii and Konev, 1965).

Two features of the fluorescence of aqueous solutions of tryptophan stand out: firstly, the absence of structure in the fluorescence spectrum, whereas the absorption spectrum and the phosphorescence spectrum, which are associated with the same systems of vibrational levels, are distinctly structured; secondly, the strong Stokes' shift (of around 70 mμ) of the fluorescence spectrum.

Since for the tryptophan molecule, as for indole, which belongs to the point symmetry group C_s, all the electronic-vibrational transitions are allowed, we can assume that the main reason for the lack of structure of the fluorescence band of tryptophan in aqueous solutions is its strong tendency to interact with the surrounding

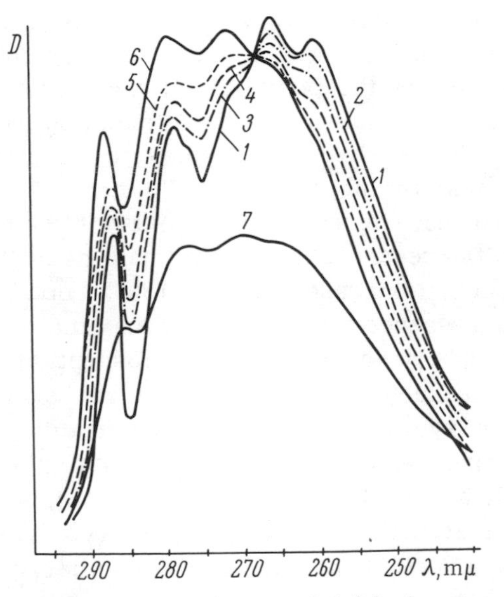

Fig. 3. Absorption spectra of indole in water (7) and in mixtures of n-hexane and n-butanol (1 to 6). Concentrations of n-butanol (in vol.%), respectively: (1) 0; (2) 1; (3) 3; (4) 5; (5) 20; and (6) 100. (Data obtained by the author in collaboration with E. A. Chernitskii.)

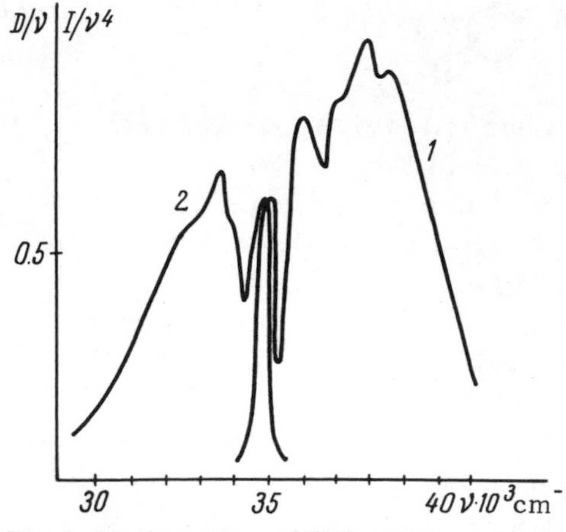

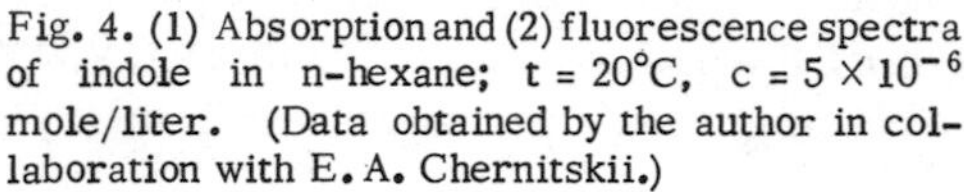
Fig. 4. (1) Absorption and (2) fluorescence spectra of indole in n-hexane; t = 20°C, c = 5×10^{-6} mole/liter. (Data obtained by the author in collaboration with E. A. Chernitskii.)

medium in the excited state. Hence, for the elucidation of the molecular structure it was of interest to examine first of all the absorption and fluorescence spectra of tryptophan in nonpolar solvents. In view of the poor solubility of tryptophan in nonpolar solvents the main measurements were made on indole.

Figure 3 shows that the absorption spectrum of indole in n-hexane has maxima at 261, 266, 279, and 287 mμ and shoulders at 271 and 276 mμ. In contrast to Weber (1961), who observed a structureless fluorescence band with maxima at 325 mμ for indole in hexane, Chernitskii and Konev (1964), obtained a distinctly structured fluorescence spectrum with maxima at 288 and 299.5 mμ and inflections at 296 and 308 mμ. The absorption and fluorescence spectra even in such a neutral solvent as hexane (Fig. 4) do not exhibit mirror symmetry. Yet the absorption and fluorescence spectra of carbazole—a compound related to indole and tryptophan and one which has well-separated ^1L_a and ^1L_b bands (Fig. 5)—are perfectly symmetrical. The absence of such symmetry in indole, where the ^1L_a and ^1L_b bands are close together, indicates that two electronic transitions are simultaneously implicated, though not in equal measure, in the formation of the long-wave band of the absorption spectrum and fluorescence spectrum. The following experimental technique can be used to decide what elements of the fine structure should be assigned to the vibrational sublevels of the ^1L_a transition and what to the vibrational sublevels of the ^1L_a electronic level. Several authors have shown that the formation of a hydrogen bond between molecules of phenols (Coggeshall and Lang, 1948; Mizushima et al., 1955; and Nagakura and Baba, 1952), pyridazines (Brealey and Kasha, 1955), naphthols (Nagakura and Gouterman, 1957), anthrols (Suzuki and Baba, 1963), and solvent molecules is accompanied by an equivalent shift of the entire long-wave absorption band of these substances. Sometimes the shift of the absorption maximum of the molecules corresponds to the energy of hydrogen-bond formation (Brealey and Kasha, 1955). In more general form the shift of the pure electronic transition $\Delta_{\nu\text{sym}}$ corresponds to the difference in energy of formation of the hydrogen bond in the excited and ground states (Pimentel, 1957).

Chernitskii and Konev (1964) observed similar relationships for a compound related to indole, carbazole. The two vibrational maxima of the electronic transition of the long-wave absorption band of carbazole on changeover from hexane solutions to alcohol solutions, where the hydrogen of the nitrogen atom of the imino group is incorporated in a hydrogen bond, are shifted in the long-wave direction through an equal number of wavelengths per centimeter. The two maxima of the fluorescence spectrum are similarly shifted in the long-wave direction through an equal number of wavelengths per centimeter (Table 1).

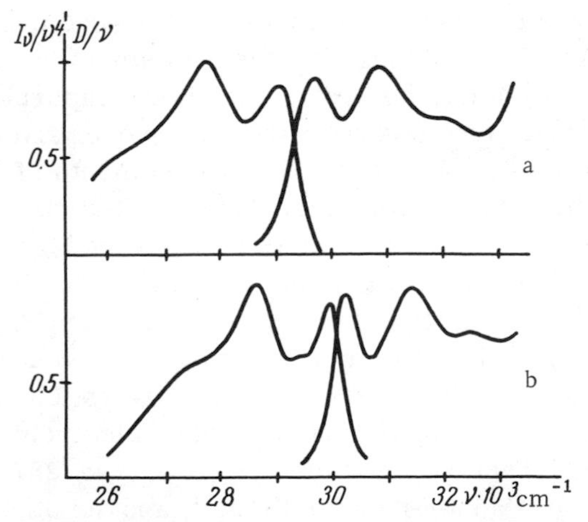

Fig. 5. Absorption and fluorescence spectra of carbazole in (a) n-butanol and (b) n-hexane (Chernitskii and Konev, 1964).

Table 1. Position of Absorption and Fluorescence Maxima of Carbazole in Hexane and Butanol

Absorption, cm^{-1}				Fluorescence, cm^{-1}			
ν_{hex}	ν_{but}	$\Delta\nu$	$\Delta\nu_{av}$	ν_{hex}	ν_{but}	$\Delta\nu$	$\Delta\nu_{av}$
30,180	29,600	580	585	30,030	29,100	930	920
31,430	30,840	590		28,740	27,830	910	

Indole, which, like carbazole, possesses an imino group, can form hydrogen bonds with the solvent. Using infrared spectroscopy, Dodd and Stephenson (1959) observed a shift of the —NH vibrations in acetone and pyridine, corresponding to hydrogen-bond formation, while Sannigrahi and Chandra (1963) determined the equilibrium constant of complexes with hydrogen bonds from the increase in absorption on the long-wave decline of the absorption band of 2-methyl indole when ether and dioxane were added to cyclohexane solutions.

When n-butanol was gradually added to hexane solutions of indole, Chernitskii and Konev (1964) observed a gradual transformation of the spectra, as in the case of carbazole. The isobestic point at 268 mμ (Fig. 3) indicates that the binary alcohol-acetone solution contains two forms of the indole molecule—a neutral form and a form bound with the alcohol by a hydrogen bond. As in the case of carbazole, all the absorption maxima are shifted in the long-wave direction, which indicates beyond question that the whole absorption band belongs to the system of π-π transitions.

From our viewpoint, however, the most important feature of the spectral shifts accompanying the changeover from a hexane to an alcohol solution lies in the fact that the elements of the structure of the long-wave absorption band are shifted through different absolute distances. The short-wave maxima undergo a greater red shift (through 700 cm^{-1}) than the long-wave maxima at 287 and 279 mμ, which are shifted through only 150 cm^{-1}. The unequal shifts of the absorption maxima due to hydrogen-bond formation have an important result: The long-wave absorption band of indole (and, hence, of tryptophan) is formed by two electronic transitions. One is $^1L_b \rightarrow A$ with vibrational maxima at 287 and 279 mμ, separated from one another by 1000 cm^{-1}; and the other, at shorter wavelength ($^1L_a \rightarrow A$) forms a series of electronic-vibrational bands beginning at 276 cm^{-1}, separated from one another by a smaller distance, 700 cm^{-1}. As we shall see, this conclusion is borne out well by polarization measurements. The different influence of the hydrogen bond on the position of the maxima corresponding to the electronic transitions $^1L_a \leftarrow A$ and $^1L_b \leftarrow A$ suggests that their moments are oriented relative to the "skeleton" of the molecule. This possibility is based on the following published information. In anthracene and naphthalene derivatives the introduction of substituents onto the molecule has a greater effect on electronic transitions for which the direction of the moment passes through the substituent (Jones, 1947; and Baba and Suzuki, 1962). The spectra are affected in this way not only by substituents, but also by hydrogen bonds (Suzuki and Baba, 1963).

The same relationships are observed in a compound related to tryptophan and indole, i.e., carbazole. Replacement of the hydrogen of the imino group with various radicals leads to a pronounced shift of the ^1L_a band, the oscillator of which is oriented along the short axis of the molecule, i.e., passes through the nitrogen atom (Platt, 1951; and Schütt and Zimmermann, 1963). On the other hand, substituents have little effect on the ^1L_b transition, which is associated with the long axis of the molecule (Omel'chenko, Pushkareva, and Bogomolov, 1957). The same applies to the effect of the hydrogen bond on the absorption bands due to the ^1L_a and ^1L_b transitions. When hydrogen bonds are formed between the OH groups of the alcohol and NH of the carbazole, ^1L_a is shifted through 580 cm^{-1}, whereas ^1L_b, the transition moment of which does not pass through the imino group, is shifted through only 370 cm^{-1} (Chernitskii and Konev, 1964). These relationships can be used to determine the orientation of the absorption oscillators of indole and tryptophan: A greater spectral shift of the short-wave maxima (^1L_a) indicates that the corresponding oscillators are oriented along the short axis of the molecule. The slightly displaced long-wave maxima of indole (279 and 287 mμ) correspond to the electronic transition $^1L_b \leftarrow A$, which is oriented along the long axis of the molecule.

We shall now consider the effect of polar solvents on the fluorescence spectrum of indole. The greater shift in the long-wave direction of the vibrational maxima of the ^1L_a transition in comparison with the vibrational maxima of ^1L_b provides additional evidence of the complex nature of the fluorescence spectrum of indole, the formation of which involves two electronic transitions, not one as in the majority of other molecules of organic substances. When

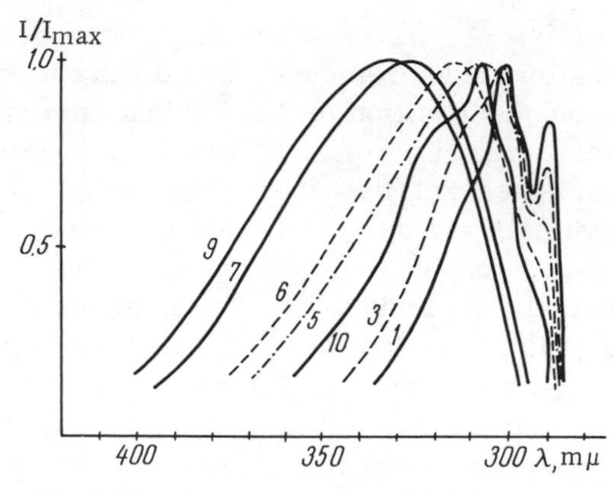

Fig. 6. Fluorescence spectra of indole in water (9) and in mixtures of n-hexane and n-butanol (3, 5, 6, and 7) at t = 20°C and in n-butanol at t = -196°C. For curves 1, 3, 5, 6, and 7 the concentration of n-butanol (in volume percentage) was: (1) 0; (3) 0.7; (5) 2.5; (6) 5; and (7) 100. (Data obtained by the author in collaboration with E. A. Chernitskii.)

n-butanol is added to a hexane solution of indole, the maximum at 268 mμ is greatly reduced in intensity and is shifted only slightly in the long-wave direction (Fig. 6). The maxima at 299.5 and 308 mμ behave differently; they are shifted farther in the long-wave direction. In butanol at 77°K the maximum at 299.5 mμ is shifted through 8 to 8.5 mμ in the long-wave direction. In indole-propionic acid (deaminated tryptophan) this shift is even more pronounced (14 mμ). Returning to the question of symmetry of the absorption and fluorescence spectra of indole in hexane solution and considering the above hypothesis that the elements of their structure are due to different electronic transitions, we can assume that the absence of symmetry of these spectra reflects the different relative roles of the two electronic transitions in the formation of the fluorescence spectrum and the different value of the Stokes' shift for each of them. Nevertheless, for several elements of the structure of the absorption and fluorescence spectra there is good partial correspondence. The fluorescence band at 288 mμ, which coincides almost resonantly with the absorption maximum at 287 mμ, must be ascribed to the ^1L_b transition. These bands include the 0-0 transition of the ^1L_b band. This electronic transition also gives rise to the maximum at 279 mμ (33,800 cm^{-1}) in absorption and the shoulder at 296 mμ (35,800 cm^{-1}) in fluorescence. In accordance with the hypothesis, the vibrational sublevel at 33,800 cm^{-1} in absorption and the vibrational sublevel at 35,800 cm^{-1} in fluorescence are at an equal energy distance from their 0-0 transition at 287.5 mμ (34,800 cm^{-1}): 1000 cm^{-1} for both absorption and fluorescence. Thus, the ^1L_b fluorescence band shows an insignificant Stokes' shift. The ^1L_a fluorescence band shows a greater shift. The frequencies of the main fluorescence maximum at 299.5 mμ (33,400 cm^{-1}) and the inflection at 308 mμ (32,600 cm^{-1}) can be compared with those of the absorption maxima at 276 mμ (36,200 cm^{-1}) and 271 mμ (36,800 cm^{-1}). These elements are separated in the absorption spectrum and fluorescence spectrum by the same frequency interval of 800 cm^{-1}. According to the rule of symmetry of absorption and fluorescence spectra, the 0-0 transition of the ^1L_a band is situated at 34,800 cm^{-1}, i.e., it practically coincides with the pure electronic transition of the ^1L_b band (34,820 cm^{-1}).

More distinctly structured quasi-line spectra can be obtained for indole in cyclohexane at 77°K. In this case the half width of the most intense lines does not exceed 120 cm^{-1} (Kembrovskii, Bobrovich, and Konev). An equally distinct structure is observed in the excitation (absorption) spectrum, which agrees well with the absorption spectrum of indole vapor (Hollas, 1963). We show the spectra taken from our work in Fig. 7 (Kembrovskii, Bobrovich, and Konev). The figure shows that the vibrational frequencies in the fluorescence spectrum can be divided into two series, one of which, with vibrational frequencies 610, 760, and 1050 cm^{-1},

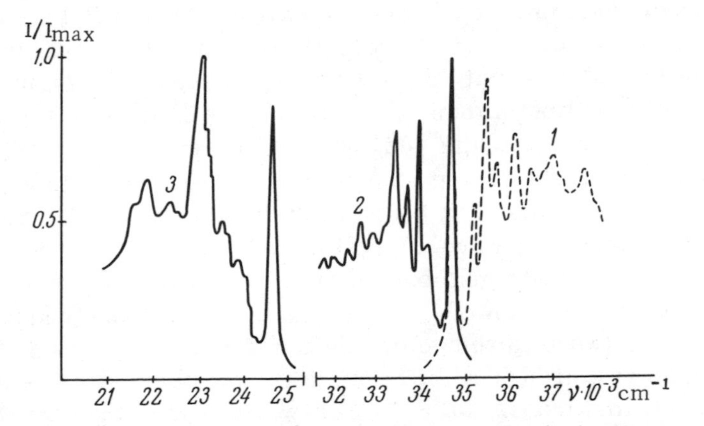

Fig. 7. Indole in cyclohexane, t = -196°C: (1) Fluorescence excitation spectra; (2) fluorescence spectrum; (3) phosphorescence spectrum. (Data obtained by the author in collaboration with V. P. Bobrovich and G. S. Kembrovskii.)

coinciding with the Raman and infrared absorption frequencies (Kohlrausch and Seka, 1938), corresponds to the $^1L_b \rightarrow A$ transition with the 0-0 frequency at 34,730 cm^{-1}. The second series with the same vibrational frequencies is revealed by a count of vibrational quanta from the 33,390 cm^{-1} band, like the 0-0 transition of ^1L_a in emission. The 0-0 band of the $^1L_a \leftarrow A$ transition in absorption does not appear, owing to the Frank-Condon principle.

The doubling of the most intense vibrational frequencies in the fluorescence spectrum indicates the complex nature of the fluorescence spectrum, which consists of ^1L_a and ^1L_b frequencies. Although structured spectra cannot be obtained for tryptophan itself, tryptophan residues contained in crystals of the protein amylase show both ^1L_a and ^1L_b components in the fluorescence spectrum at liquid nitrogen temperature.

The above account shows that the fluorescence spectra of the usual molecule of indole or tryptophan in a strongly polar (aqueous) environment are not at all like the fluorescence spectra of the "isolated" molecules under conditions in which the environment causes the least possible perturbation of the molecular energy levels. What is the fundamental nature of the interactions of the tryptophan molecule with the environment which lead to such pronounced distortions of the energetics of the molecule? Firstly, for organic molecules which cannot form hydrogen bonds, the positions of the fluorescence maxima in hexane solutions and n-butanol at low temperature are usually the same. In contrast to this, for substances capable of forming hydrogen bonds, such as naphthalene or anthracene derivatives (Suzuki and Baba, 1963) or carbazole (Chernitskii and Konev, 1964), hydrogen-bond formation is accompanied by a shift of the total fluorescence spectrum in the long-wave direction without alteration of its shape. In the case of indole or tryptophan, the long-wave shift is accompanied by an additional effect, the disappearance of the short-wave fluorescence maxima. Hence, the first reason for the shift of the fluorescence spectra in the long-wave direction in any polar medium is the formation of a hydrogen bond between the imino group and the solvent molecules. However, since the hydrogen bond is apparently formed with two different electronic excited molecules and shifts their spectra through a different distance, the result is that the 0-0 ^1L_a transition is in a longer-wave region than the 0-0 ^1L_b transition. The energy distance between them is increased so much that the indole and tryptophan molecules cease to be "peculiar" and begin to luminesce, like other substances, only from the lowest electronic level. Thus, in addition to causing the long-wave shift of the fluorescence spectrum, the hydrogen bond weakens the intensity of the ^1L_b fluorescence.

However, hydrogen-bond formation does not exhaust the possible interactions of tryptophan and indole molecules with the solvent. The addition to n-hexane solutions of small amounts of butanol (tenths of volume percentage), which cannot alter the macroscopic dielectric properties of the medium, leads to a progressive shift of the fluorescence spectrum in the long-wave direction. The shift of the fluorescence spectra on changeover from hexane to alcohol solutions is from 289.5 to 340 mμ for indole and from 305 to 345 mμ for indolepropionic acid.

Besides hydrogen-bond formation, solvent relaxation effects are also implicated in this shift (Cherkasov, 1960 and 1962). In the ground state the forces of intermolecular interaction cause some orientation of the polar molecules (alcohol) around tryptophan or indole. Conversion of the molecule to the singlet excited state alters the magnitude and direction of the dipole moment. The change in the physical properties of the molecule is accompanied by a change in the strength and geometry of the interaction with the nearest solvent molecules and by corresponding changes in the equilibrium orientation. When the relaxation time is commensurate with the lifetime of the excited state (of the order of 10^{-8} to 10^{-9} sec) there is a whole set of excited molecules caught at the moment of emission in different stages of relaxation or, in quite general form, in different states of interaction with the nearest solvent molecules. Solvent relaxation processes, in addition to broadening the emission band, will result in the long-wave part of the spectrum's being formed by the luminescence of molecules with the greatest lifetime, during which relaxational effects can be completed. In fact, the addition of quenchers to hexane solutions of indole and indolepropionic acid does not lead to a change in the shape of the fluorescence spectra, but a short-wave shift of the fluorescence spectrum occurs in the same solutions containing a small admixture of butanol. All this indicates that various centers are implicated in fluorescence in solvents containing polar molecules, and the long-wave part of the spectrum is due to the luminescence of molecules with the longest lifetime. The relaxational nature of the long-wave shift and the broadening and smearing of the structure of the fluorescence spectrum in the presence of small amounts of polar molecules are revealed in the fact that, when the possibility of relaxation is ruled out by an increase in viscosity or a reduction in temperature (cooling of alcohol solutions to freezing point or lower), there is a short-wave shift of the spectrum and a vibrational structure appears. Similarly, an increase in the viscosity of glycerol solutions by temperature reduction leads to a short-wave shift of the fluorescence spectra.

Thus, the specific interactions (in the terminology of Neporent and Bakhshiev, 1960) taking place between molecules of the dissolved substance and one or several of the surrounding particles of the medium are manifested in the case of indole and its derivatives in the formation of hydrogen bonds and orientation solvation with relaxation of the solvate shell relative to the excited molecule. These microinteractions are very strong and lead to three results: (1) a long-wave shift of the fluorescence spectra; (2) their broadening; (3) loss of structure.

It is quite understandable that, in addition to specific (individual) interactions, the molecules of indole and tryptophan, like other substances, can take part in universal interactions resulting from the effect of the whole assemblage of surrounding molecules on the dissolved molecule. The dielectric properties of solvents are mainly responsible for the shift of the fluorescence maxima from 330 to 342 mμ for indole, from 345 to 360 mμ for indolepropionic acid, and from 338 to 350 mμ for tryptophan on changeover from alcohol solutions to aqueous solutions (Fig. 8). The universal effect of the solvent and its dielectric properties on the position of the fluorescence maximum of tryptophan was shown by Teale (1960) for the case of water-dioxane solutions of glycyltryptophan. The position of the maximum was shifted in the short-wave direction in correspondence with the reduction of the dielectric constant of the binary solvent.

In a pure dioxane solution, tryptophan has a maximum at 329 mμ, as against 350 mμ in an aqueous solution. The position of the fluorescence maximum and the shape of the spectrum in actual conditions are presumably determined by the effect of the three main types of interaction: the universal type (the effect of the integral field of the solvent on the dipole moments of the tryptophan molecule in the excited state) and two specific types (the hydrogen-bond effect and solvent relaxation effect). The greater effect of interaction with the environment on the position of the levels of the excited state (greater shifts in fluorescence spectra than in

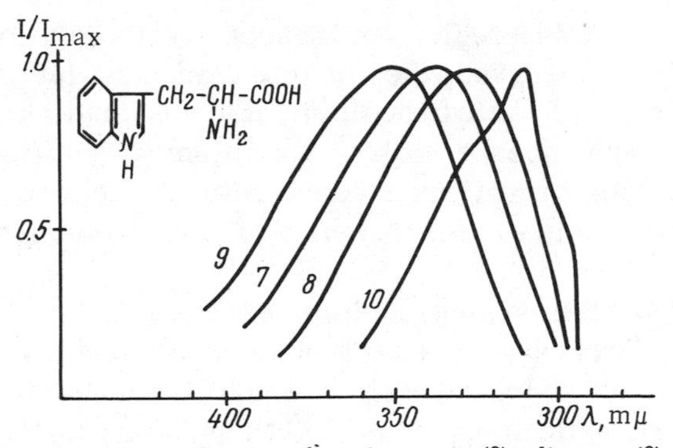

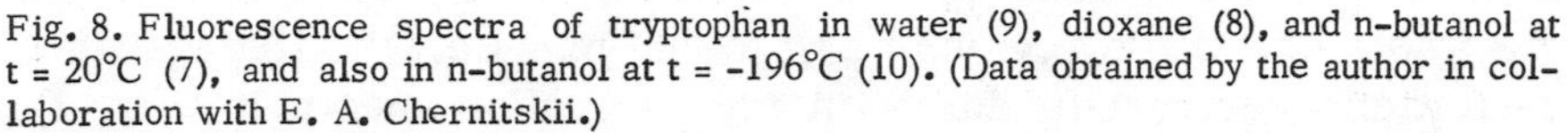
Fig. 8. Fluorescence spectra of tryptophan in water (9), dioxane (8), and n-butanol at t = 20°C (7), and also in n-butanol at t = -196°C (10). (Data obtained by the author in collaboration with E. A. Chernitskii.)

absorption spectra) than on the ground state is due to the fact that the dipole moment of the excited molecule becomes much greater, and, subsequently, there is an increase in the strength of interaction with the environment (interaction of constant or induced charges). Probably the most important of all the causes for the long-wave shift of the fluorescence spectra of indole and its derivatives is the effect of the dielectric constant of the medium. This is indicated by the shift in the maximum of the fluorescence spectrum from 322 to 380 mμ on changeover from a hexane to an aqueous solution in the case of 1,3-dimethyl indole, which cannot form hydrogen bonds with the solvent.

In conclusion we shall consider the frequently discussed problem of the fluorescence of tryptophan in the visible region of the spectrum. Steele and Szent-Györgyi (1958), Isenberg and Szent-Györgyi (1958), and Fujimori (1960) found that concentrated solutions of tryptophan (5×10^{-3} to 10^{-2} mole/liter) in glucose at low temperatures showed, in addition to the usual fluorescence (325 mμ and phosphorescence with three maxima in the region of 400 to 500 mμ), an unusual fluorescence at 450 mμ with a corresponding phosphorescence at 500 mμ, which were excited with low efficiency by light with $\lambda = 365$ mμ. If we exclude the possibility that this luminescence is due to impurities, then we might presume that it is due, e.g., to $n \rightarrow \pi$ transitions in the tryptophan molecules, which interact strongly with one another (and, possibly, with additives) via the hydrogen of the imino group. This is all the more likely in view of the fact that, according to Steele and Szent-Györgyi's observations, traces of an extremely weak absorption maximum can be detected in this region in concentrated solutions and that Fujimori (1960) observed the disappearance of luminescence at 450 and 500 mμ in 8-M hydrochloric acid. Chernitskii and I later observed phosphorescence with a maximum at 500 mμ for sugar crystals of tryptophan at room temperature. Lehrer and Fasman (1964) reported excimer fluorescence of polytryptophan at 470 mμ.

Quantum Yield of Fluorescence and Its Dependence on the Structure of the Tryptophan Molecule and External Conditions

According to Teale and Weber (1957), the quantum yield of the fluorescence of natural aqueous solutions of tryptophan is 0.2 for the whole excitation region of 210 to 300 mμ, i.e., the fluorescence excitation spectra coincide with the absorption spectrum (Fig. 1). The quantum yield of fluorescence is sensitive to the ionization state of the functional groups. White (1959) investigated the dependence of the quantum yield of tryptophan fluorescence on the pH of the medium and found a distinct relationship, shown in Fig. 9, which reflects the ionization processes in this amino acid. Similar titration curves for tryptophan fluorescence were later obtained by many authors (Weber, 1961; Vladimirov and Li Chin-kuo, 1962; Cowgill, 1963A; and Konev and Chernitskii, 1964).

Strongly quenching groups in the tryptophan molecule are the nonionized carboxyl group COOH, the amino group after it has

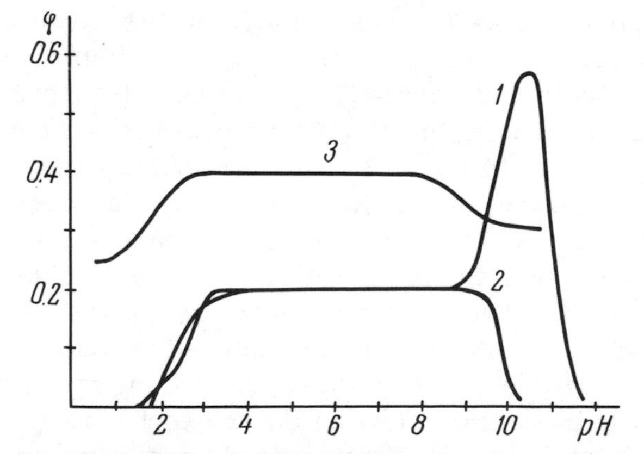

Fig. 9. Quantum yield of fluorescence of tryptophan (1), tyrosine (2), and phenylalanine (3) as a function of the pH of the medium. The values of the yield of phenylalanine are increased ten times [1 and 2 from White (1959); 3 from Feitelson (1964)].

Table 2. Effect of Substituents on Quantum Yield of Fluorescence of Tryptophan Derivatives (Cowgill, 1963)

$CH_2-CH(R_2)-R_1$ (indole, N–H)

Compound	R_1	R_2	Q
Tryptophan	$-COO^-$	$-NH_2$	0.51
N-acetyl tryptophan	$-COO^-$	$-NH-CO-CH_3$	0.28
Tryptophan amide	$-CO-NH_2$	$-NH_2$	0.28
Tryptophan	$-COO^-$	$-\overset{+}{N}H_3$	0.20
N-acetyl tryptophan	$-COO^-$	$-NH-CO-CH_3$	0.128
Tryptophan	$-COOH$	$-\overset{+}{N}H_3$	0.085
Methyl ester of acetyl tryptophan	$-COOCH_3$	$-NH-CO-CH_3$	0.080
Ethyl ester of tryptophan	$-COOC_2H_5$	$-NH_2$	0.076
Ethyl ester of tryptophan	$-COOC_2H_5$	$-\overset{+}{N}H_3$	0.032
Glycyltryptophan	$-COO^-$	$-NH-CO-CH_2-NH_2$	0.095
Glycyltryptophan	$-COO^-$	$-NH-CO-CH_2-\overset{+}{N}H_3$	0.057
Gylcyltryptophan	$-COOH$	$-NH-CO-CH_2-\overset{+}{N}H_3$	0.022

acquired a proton, and the imino group after it has lost a proton. The most efficient fluorescence quenchers are the first and third groups. It is known that tryptophan, like any amino acid, is an ampholyte, capable of dissociating as an acid or as a base

$$^{+}H_3N-R-COO^{-}+H^{+} \overset{k_1}{\rightleftarrows} {}_3H^{+}N-R-COOH$$

$$\overset{+}{_3}HN-R-COO^{-} \overset{k_2}{\rightleftarrows} {}_2HN-R-COO^{-}+H^{+}$$

$$K_1 = \frac{[\overset{+}{_3}HN-R-COOH]}{[\overset{+}{_3}HN-R-COO^{-}]\,H^{+}} \qquad K_2 = \frac{[{}_2HN-R-COO^{-}]\,H^{+}}{\overset{+}{_3}HN-R-COO^{-}}$$

pK_1, or the pH at which the concentrations of $\overset{+}{_3}HN-R-COO^-$ and $\overset{+}{_3}HN-R-COOH$ are equal, i.e., the pH at which half of the molecules have charged carboxyl group and half have a neutral group, is 2.3 for tryptophan; pK_2 of the amino group in an alkaline medium is 9.38. Hence, it is quite understandable that changes in quantum yields due to ionization of the side groups of tryptophan should occur close to these two pH values. In fact, with increase in the pH values, beginning from the most acid solutions, there is a two-step increase in quantum yield, reflecting the course of dissociation of the side groups. In an acid medium the luminescence quenching is most efficient, since in it the substituent groups are in the state COOH and $-NH_3^+$. The increase in quantum yield with increase in pH from pF = 2.3 (the point corresponding to the center of the yield growth curve) corresponds exactly with the amperometrically found pK for dissociation of the carboxyl group (2.3).

According to Cowgill's data (1963) tryptophan with a nonionized COOH group and a protonized amino group has a low quantum yield of fluorescence, 0.085 (Table 2).

The mechanism of the acid quenching of tryptophan fluorescence is definitely indicated not only by the coincidence of the titration curves for the carboxyl group and the "fluorescence," but also by the absence of quenching—a shift of the pK of fluorescence to 0.5—in compounds without a carboxyl group (indole) or in compounds where this group is in a chemically bound state (the methyl ester of N-acetyl tryptophan). In the pH range of 4 to 8, all the tryptophan molecules in solution are in the state $_3^+HN-R-COO^-$, for which there is a constant and pH-independent quantum yield of 0.2 (Teale and Weber, 1957; White, 1959; and Cowgill, 1963). In the more alkaline region, the protonized amino group begins to change into a neutral group ($_3^+HN-R-COO^- \rightarrow {}_2HN-R-COO^-$), and this is accompanied by a further increase in quantum yield. The pF of this transition coincides with the pK of dissociation of the amino group, although there is a slight shift in the alkaline direction: pF = 9.4 (White, 1959), 9.5 (Cowgill, 1963), and 9.4 (Vladimirov and Li Chin-kuo, 1962), whereas pK = 9.38 (French and Edsall, 1945). Tryptophan in the ionic form $_2HN-R-COO^-$ has a quantum yield of 0.51 (Cowgill, 1963). This quantum yield, the greatest for aqueous solutions, is observed at pH 10.9. That the second step of the quantum-yield growth curve in the pH region of 8 to 11 is due to dissociation of the amino group is also indicated by the constancy of the quantum yield in compounds which do not have an amino group—N-methyl indole (White, 1959), indole, and indolepropionic acid (Fig. 10) (Konev and Chernitskii, 1964)—or in compounds where the hydrogen of the amino group is replaced by another radical—the methyl ester of N-acetyl tryptophan (Cowgill, 1963). The maximum possible values of the quantum yield (0.51) are found

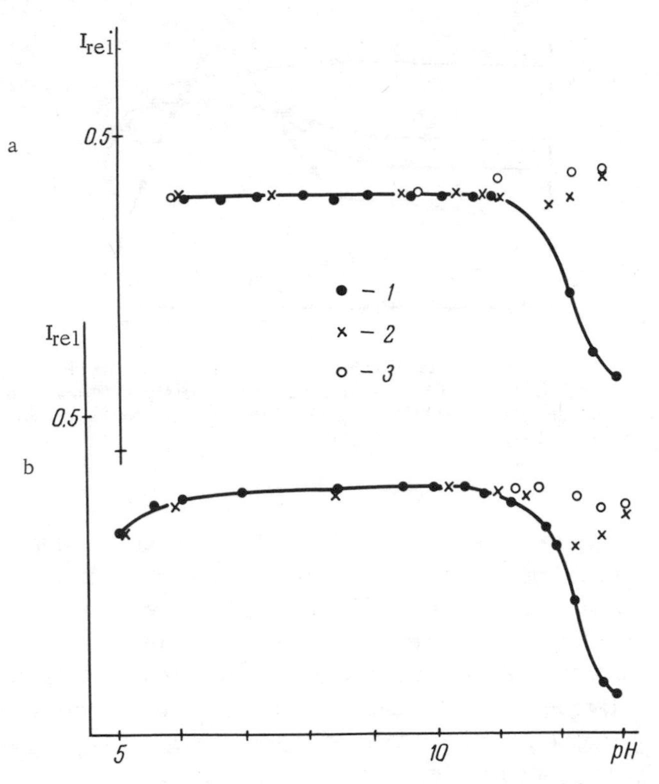

Fig. 10. Quantum yield of fluorescence of (a) indole and (b) indolepropionic acid as a function of the pH of the medium: (1) in aqueous solution, (2) immediately after the addition of 1% formaldehyde, and (3) in 1% formaldehyde 1 hr after the addition (Konev and Chernitskii, 1964).

in tryptophan in a wide pH range (4 to 12) if the amino group is blocked by the addition of formaldehyde in an alkaline medium (Fig. 11), where chemical bonds of the type $R-N = CH_2$ or $R-NH-CH_2OH$ are formed (Konev and Chernitskii, 1964). Formaldehyde has no effect on the quantum yields of indole and indolepropionic acid, which do not contain an amino group (except in the pH region of 11 to 13, where blocking of the imino group becomes effective).

In the literature there are two very close points of view on the mechanism of the changes in the quantum yield due to ionization of the carboxyl and amino groups of tryptophan. According to White (1959), whose opinion is shared by Weber (1961), quenching by the COOH and NH_2 groups has a common mechanism, i.e., quenching by the proton belonging to these groups, which is attracted to the

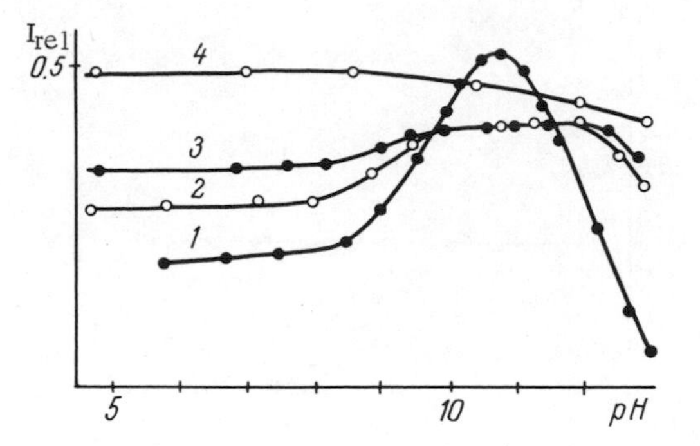

Fig. 11. Quantum yield of tryptophan fluorescence as a function of the pH of the medium: (1) in aqueous solution, (2) in 1% formaldehyde immediately after addition, (3) in 2% formaldehyde immediately after addition, and (4) in 1% formaldehyde a day after addition (Konev and Chernitskii, 1964).

indole ring in the excited state. Cowgill (1963) starts out from the electronegativity of the side groups, i.e., from their ability to attract electrons of the indole ring. With increase in the number of neutral methyl groups between the indole ring and the electronegative substituent, the magnitude of its effect on the quantum yield gradually decreases: the quantum yields of indole—COO^-, indole—CH_2—COO^-, and indole—CH_2—$CH(NH_2)$—COO^- are 0.24, 0.38, and 0.51, respectively.

The role of the third functional group of tryptophan, the imino group, is manifested only in a strongly alkaline medium, beginning at pH 11. Quenching of the fluorescence in a strongly alkaline medium is due to the presence of this functional group. The role of the mobile hydrogen of the imino group is confirmed by the absence of quenching in a strongly alkaline medium in the case of N-methyl indole (White, 1959) and the methyl ester of N-acetyl tryptophan (Cowgill, 1963). There is no decrease in the intensity of the fluorescence of indole and tryptophan when the imino group is blocked by formaldehyde (Fig. 12) (Konev and Chernitskii, 1964).

White suggested the following mechanism to explain alkaline quenching:

$$RNH + h\nu \rightarrow RNH^*$$

$$RNH^* + OH^- \rightarrow RN^{-*} + OH_2$$

$$RN^{-*} \rightarrow RN^- \quad \text{(radiation transition)}$$

$$RN^- + OH_2 \rightarrow RNH + OH^-$$

Following Boaz and Rollefson (1950), who showed the radiationless nature of the $R^*N^- \rightarrow RN^-$ transition in α-naphthylamine, White regards quenching by hydroxyl ions as a transfer of the H ion from the nitrogen atom of the excited tryptophan molecule to the ion of the OH solvent. White's point of view is shared by Weber (1961), Vladimirov and Li Chin-kuo (1962), and Udenfriend (1962). This scheme, however, requires some amendment.

The first premise in White's scheme—the nondissociability of the imino group in the ground unexcited state—is not in question. The absence of ionized forms of the indole ring in the unexcited state is confirmed by the coincidence of the alkaline titration curves for indole and water, indicating the absence of the reaction

$$RNH + OH^- \rightarrow RN^- + H_2O$$

which would have required the use of additional amounts of alkali (Chernitskii and Konev, 1965).

The situation is different for the second premise of White's scheme, the inability of the ionized form of the indole ring to fluoresce. My measurements in collaboration with Chernitskii (1965) showed that, in a 4% NaOH solution, indole and tryptophan exhibit fluorescence with a quantum yield approximately one-and-

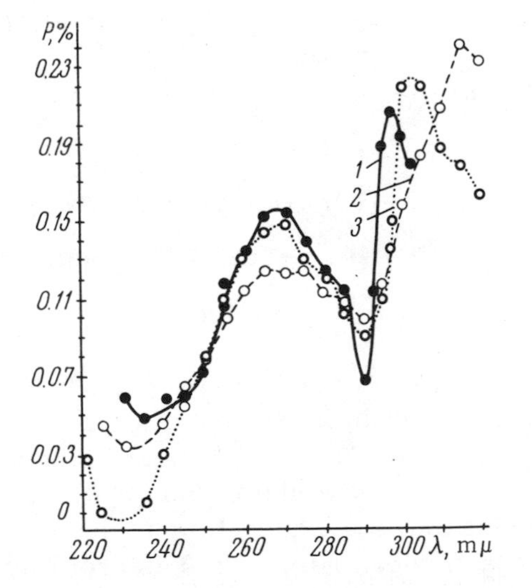

Fig. 12. Fluorescence polarization spectra of indole (1), N-methyl indole (2), and tryptophan (3) (Weber, 1960).

a-half orders lower than in neutral solvents and with maxima at 405 and 420 mμ, respectively (Fig. 2). Thus, ionization of the imino group in the excited state does not lead to simple quenching of the luminescence, but to the appearance of molecules capable of luminescence in the longer-wave region of the spectrum. Moreover, the quantum yield of the fluorescence of these molecules has low values not because of the ease of thermal dissipation of the electronic excitation energy within the indole ring as White suggests, but because of kinetic quenching of the luminescence of the ionized molecule by collisions with hydroxyl ions. At 77°K the fluorescence of alkaline indole, which is shifted considerably towards longer wavelengths in comparison with a neutral solution (370 mμ as against 320 mμ), nevertheless has the initial quantum yield values. Hence, if kinetic collisions are excluded, photodissociation of the imino group occurs (the spectrum is shifted), although quenching ceases. This allows the separation in time of photoionization and quenching if it is assumed that tryptophan and indole molecules ionized in the imino group can themselves luminesce but, in the presence of an excess of alkali, are quenched by collision with hydroxyl ions. Thus, in the more accurate form the scheme of fluorescence quenching in a strongly alkaline medium due to photoionization of the imino group of indole and tryptophan, followed by quenching of the luminescence by collision with hydroxyl ions, will be as follows (Chernitskii and Konev, 1965):

$$RHN + h\nu_1 \rightarrow RNH^*$$

$$RNH^* + OH^- \rightarrow RN^{-*} + H_2O$$

$$\begin{cases} RN^{-*} \rightarrow RN^- + h\nu_2 \text{ (fluorescence)} \\ RN^{-*} + OH^- \rightarrow RN^- + OH^- \text{ (quenching)} \end{cases}$$

$$RN^- + H_2O \rightarrow RNH + OH^-$$

The fluorescence of tryptophan is quenched by heating (Gally and Edelman, 1962) and by additions of oxygen (Barenboim, 1963; and Barenboim and Domanskii, 1963), potassium iodide (Vladimirov and Li Chin-kuo, 1962), and lithium bromide (Konev and Chernitskii, 1964A). In the last two cases quenching is due to the facilitation of conversion to the triplet state. This is probably also the reason for the weak fluorescence of indole in hexane solutions at low temperatures, where intense phosphorescence occurs.

Polarization Spectra of Tryptophan Fluorescence

Absorption Polarization Spectra of Fluorescence. Absorption polarization spectra of the fluorescence of tryptophan and indole were obtained by Weber in 1960. The spectra were recorded in propylene glycol at -70°C, and the fluorescence was excited by unpolarized light. Hence, to obtain the usually adopted values of the polarization P, we have to multiply Weber's figures by a factor $2P_u/(P_u + 1)$, where P_u is the polarization produced on excitation by unpolarized light. According to Weber's data, indole, N-methyl indole, and tryptophan have similar polarization spectra with two minima at 232 ± 2 and 290 mμ and two maxima at 265 to 270 and 300 to 305 mμ (Fig. 12).

The low absolute values of the polarization over the whole long-wave absorption band, except for its extreme long-wave end, which are not characteristic of molecules with low symmetry, are well approximated, as is shown by a linear oscillator. The most important point, the noncoincidence of the polarization spectrum and the absorption spectrum, led Weber to postulate that the long-wave absorption band is of a complex nature and consists of two electron transition moments oriented at right angles. Weber noted that only one of the two electronic levels is implicated in emission; the second merely transfers absorbed energy to the first. This system of absorption oscillators gives a clear picture of the experimentally observed main details and special features of the polarization spectrum. The high values of the polarization (35 to 40% in the region of 300 to 310 mμ) correspond to the absorption of light by only the ^1L_a oscillator; the minimum at 290 mμ corresponds to the absorption maximum of the "negative" ^1L_b oscillator, which reduces the positive polarization of the ^1L_a oscillator; the maximum at 265 to 270 mμ corresponds to the ^1L_a absorption maximum, with the result that the relative contribution of the ^1L_b oscillator to absorption at this point of the spectrum is reduced and a relatively large positive polarization is observed, etc. However, Weber's investigations give no evidence that the luminescence is due to only one kind of luminescent molecule. It is obvious that the presence of an impurity or a different ionic form of tryptophan with an absorption band at 290 mμ could lead to the appearance of a minimum in the fluorescence polarization spectrum in this region. Hence, evidence showing that the two absorption oscillators belong to the same luminescent molecule is of primary importance.

Chernitskii, Konev, and Bobrovich (1963) showed that in a solid

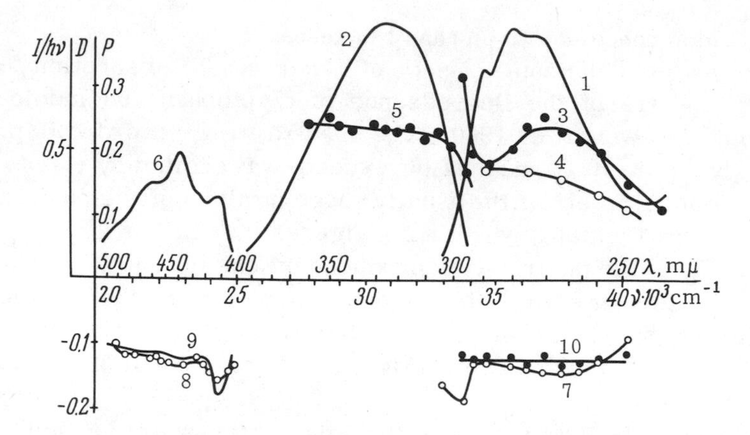

Fig. 13. Spectra of DL-tryptophan in polyvinyl alcohol film at 20°C: (1) absorption spectrum, (2) fluorescence spectrum, (3) absorption polarization spectrum of fluorescence (recorded at 330 mμ), (4) the same (recorded at 295 mμ), (5) emission polarization spectrum of fluorescence (excitation 265 mμ and t = -196°C), (6) phosphorescence spectrum, (7) absorption polarization spectrum of phosphorescence (recorded at 460 mμ), (8) emission polarization spectrum of phosphorescence (excitation 265 mμ), (9) spontaneous emission polarization spectrum of phosphorescence (excitation 265 mμ), (10) spontaneous absorption polarization spectrum of phosphorescence (recorded at 460 mμ) (Chernitskii, Konev, and Bobrovich, 1963).

solution of polyvinyl alcohol the fluorescence spectrum of tryptophan in the 250- to 300-mμ region is independent of the wavelength of the exciting light, and yet the polarization spectrum has the usual shape, with a dip in the region of 289 to 290 mμ (Fig. 13).

The characteristic shape of the polarization spectrum is manifested in almost unaltered form in the case of viscous solvents of different chemical nature: glycerol (Konev, Katibnikov, and Lyskova, 1964), sucrose crystals (Chernitskii, 1964), and polyvinyl alcohol (Sevchenko, Konev, and Katibnikov, 1963; and Chernitskii, Konev, and Bobrovich, 1963). The authors obtained similar spectra for tryptophan and for indole.

The second piece of evidence indicating that the two absorption oscillators belong to the same molecules is given by experimental investigations of the polarization spectra of tryptophan fluorescence in relation to increasing concentration in polyvinyl alcohol. These spectra showed the same concentration depolarization at 289 mμ as in the remaining part of the absorption polarization spectrum in the range of 240 to 300 mμ (Fig. 14) (Konev, Katibnikov, and Lyskova, 1962 and 1964; and Chernitskii, Konev, and Bobrovich, 1963). The agreement of concentration depolarization at different points in the spectrum indicates that at all excitation points we are dealing with identical molecules in the same absolute concentration.

Similar effects are observed in the case of concentration depolarization of phosphorescence. The existence of the ^1L_b oscillator in the long-wave absorption band is particularly well demonstrated in experiments with oriented tryptophan (Sevchenko, Konev, and Katibnikov, 1963; and Chernitskii, Konev, and Bobrovich, 1963). Orientation of the molecules can be achieved by mechanical stretching of tryptophan-activated polyvinyl alcohol films. We use the following experimental procedure: The plane of the film is at an angle of 45° to the direction of propagation of the exciting beam, so that the ^1L_a oscillator is vertical. The fluorescence is excited by the horizontal component of the linearly polarized light. The polarization of the fluorescence is measured at its main maximum at an angle of 90° to the direction of propagation of the exciting light (Fig. 15). Such an arrangement can provide a direct answer to the questions of whether the ^1L_b oscillator exists and what kind of absorption spectrum is formed by it.

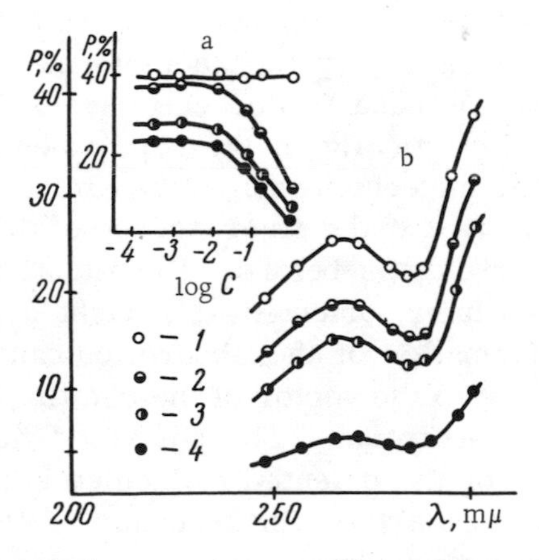

Fig. 14. Concentration quenching and concentration depolarization of fluorescence of tryptophan in a solid film of polyvinyl alcohol: (a) relative quantum yield of tryptophan fluorescence as a function of concentration (1) and concentration depolarization of tryptophan fluorescence on excitation by wavelengths 302, 265, and 289 mμ (2, 3, and 4), respectively; (b) fluorescence polarization spectra of tryptophan in concentrations of (moles per liter): (1) 10^{-2}, (2) 7×10^{-2}, (3) 10^{-1}, and (4) 1 (Konev, Katibnikov, and Lyskova, 1964).

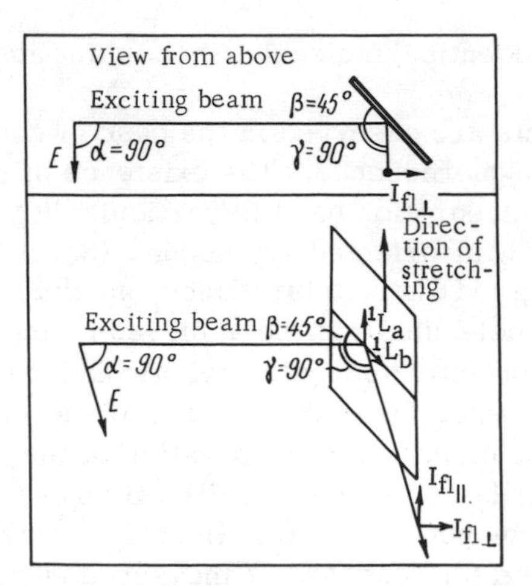

Fig. 15. Experimental setup for polarization measurement of absorption spectrum due to ^1L_b oscillator.

In fact, in such an experimental arrangement the unoriented molecules, irrespective of their oscillator nature, will give completely depolarized fluorescence over the whole absorption spectrum, as curve 7 (Fig. 16) shows. We shall now consider what polarization the oriented tryptophan molecules will give. If the long-wave absorption band is due entirely to the ^1L_a oscillator, which is oriented vertically in the experimental conditions, the oriented tryptophan molecules will not be excited at all by horizontally polarized light, and the fluorescence of the film will still be completely depolarized, as before. The situation will be different if the second oscillator, oriented at a right angle to the first, is involved in the formation of the absorption band. This oscillator is at an angle of 45° to the vector of the exciting light and hence is implicated in its absorption. The polarized fluorescence due to the ^1L_a oscillator of the oriented molecules is now superimposed on the formerly depolarized fluorescence of the chaotically arranged tryptophan molecules.

This part of the fluorescence of the film may be 100% polarized, and its intensity, directly proportional to the number of quanta absorbed by the ^1L_b oscillator in relation to the total absorption by the two oscillators. In other words, the polarization spectrum obtained in this case reflects the relationship

$$P_\lambda = f\left(\frac{\varepsilon^1 L_b \lambda}{\varepsilon^1 L_a \lambda + \varepsilon^1 L_b \lambda}\right)$$

and is actually the relative absorption spectrum of the negative oscillator.

We shall now consider the results obtained from experiments. Figure 16 shows a number of polarization spectra obtained for an oriented film of tryptophan under different conditions of excitation and detection. Curve 4 in Fig. 16 shows that the polarization for the wavelength region below 300 mμ is not zero at all but has positive values which vary in a definite manner over the absorption spectrum. The very presence of nonzero polarizations definitely indicates the presence in the long-wave absorption band of the second oscillator ^1L_b, oriented at almost a right angle to the ^1L_a oscillator, since only the presence of such an oscillator enables the oriented tryptophan molecules to absorb light in this experimental arrangement. The positive values of the polarization indicate that the ^1L_b oscillator fluoresces mainly through ^1L_a and not by itself, since otherwise the polarization would be negative. The nature of the relationship $p = f(\lambda)$ shows that the absorption maximum for the ^1L_b electron transition lies at 289 mμ, i.e., where the minimum of the absorption polarization spectrum of the fluorescence lies. At 300 mμ absorption by ^1L_b is absent, and at this wavelength the

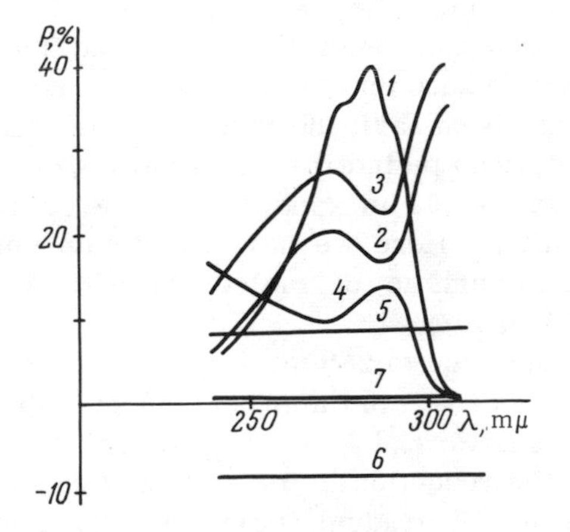

Fig. 16. (1) Absorption spectrum of tryptophan in polyvinyl alcohol; (2 and 7) relationship $P_{fl} = f(\lambda_{exc})$ for tryptophan in an isotropic film of polyvinyl alcohol excited by light polarized in vertical and horizontal directions, respectively; (3 and 4) the same as 2 and 7, but for tryptophan in an anisotropic film; (5 and 6) spontaneous polarization of tryptophan fluorescence for vertical and horizontal positions of stretched film, respectively (Sevchenko, Konev, and Katibnikov, 1963).

degree of polarization of viscous solutions of tryptophan becomes the same (35 to 40%) as for molecules with a low degree of symmetry, which can be approximated by a linear oscillator. The absorption minimum of the negative oscillator is also observed at 270 mμ, which corresponds to the main maximum in the usual polarization spectrum of an isotropic film at 270 mμ.

In several cases it is even possible to observe a vibrational structure of the ^1L_b electronic transition in an investigation of the absorption and fluorescence spectra of indole in neutral solvents and in the induction of Shpol'skii effects. It is natural that the maxima of the vibrational structure of ^1L_b should be manifested in two minima in the long-wave part of the polarization spectrum. Such an absorption polarization spectrum of fluorescence with two minima in the long-wave part of the absorption band was found for N-glycyltryptophan in a 50% propylene glycol-water mixture at -70°C by Weber (1960) and by us for indole in a sugar crystal at room temperature. The minima of the polarization spectrum lay at 286 and 293 mμ for glycyltryptophan and at 282 and 289 mμ for indole.

Thus, the absorption polarization spectra, like the data for the absorption and fluorescence spectra in different solvents, confirm the presence in the long-wave absorption band of two electronic transitions ^1L_a and ^1L_b. The electronic transition $^1L_b \leftarrow A$ forms a narrow absorption band with a maximum at 289 mμ lying inside a broader absorption band due to the electronic transition $^1L_a \leftarrow A$. In view of what has been said, the absorption polarization spectrum of the fluorescence can be interpreted as follows:

1. The region of 300 mμ and more: The same ^1L_a oscillator absorbs and emits. Here we have maximum positive values of polarization characteristic of molecules with a low degree of symmetry (p = 38 to 40%).

2. The region of 300 to 289 mμ: There is a sharp reduction in the polarization with a minimum at 289 mμ. The two mutually perpendicular oscillators ^1L_a and ^1L_b absorb but, as before, ^1L_a mainly emits. The reduction in the polarization in the short-wave direction is due to the gradual increase in the contribution of the negative oscillator to absorption. The minimum at 289 mμ corresponds to the 0-0 ^1L_b transition, where its relative contribution to the total absorption is greatest.

3. The increase in the polarization at 289 to 265 (270) mμ with a maximum at 265 to 270 mμ corresponds to the increase in

the relative absorption of the ^1L_a electronic transition, which has a maximum at 270 mμ.

4. The reduction of the polarization at 265 to 230 mμ reflects the increase in the relative absorption of the ^1L_b oscillator as a result of interaction with the ^1L_b oscillator of the absorption band at 220 mμ.

Emission Polarization Spectra of Fluorescence. The section devoted to the fluorescence spectra of tryptophan provided evidence that not only the long-wave absorption band, but the fluorescence spectrum itself is formed by the two electronic transitions $^1L_a \rightarrow A$ and $^1L_b \rightarrow A$. In addition, the absorption polarization spectra of the fluorescence indicate different spatial orientation of the ^1L_a and ^1L_b transition moments relative to one another.

Hence, it was to be expected that, in the case of monochromatic excitation, the oscillator forming the short-wave part of the fluorescence spectrum would not have the same direction as the oscillator responsible for the remaining longer-wave part of the spectrum. In other words, the formation of the fluorescence band by two different electronic transitions would be manifested in significant differences between the polarization of the short-wave portion and that of the remainder of the fluorescence band.

The shape of the fluorescence polarization spectrum of tryptophan in a solid film of polyvinyl alcohol completely confirms this prediction (Fig. 13, curve 5). In fact, the polarization retains its high and constant positive values (about 25%) over the whole fluorescence band, except for the shortest-wave part. From 315 mμ there is a gradual reduction in the degree of polarization, which attains very low values (17%) in the region of 295 mμ.

The situation is different in a polar solvent, glycerol, where the fluorescence is only from the ^1L_a level and where accordingly there is no reduction of polarization in the short-wave part of the fluorescence spectrum.

Similar emission polarization spectra of the fluorescence were obtained for indole in ethanol at 100°C (Zimmermann and Joop, 1961) and in polyvinyl alcohol at room temperature (Chernitskii, Konev, and Bobrovich, 1963).

For indole in polyvinyl alcohol there are very marked differences between the polarization of the fluorescence in the main part of the band and that in its short-wave end: The polarization falls from 25 to 10% (Fig. 17, curve 5).

The reduction in the polarization of the fluorescence would

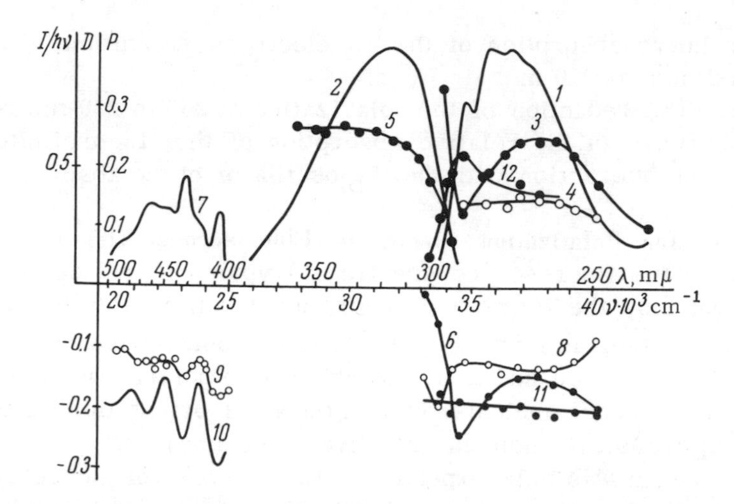

Fig. 17. Spectra of indole in film of polyvinyl alcohol at t = 20°C: (1) absorption spectrum; (2) fluorescence spectrum; (3) absorption polarization spectrum (recorded at 330 mμ); (4) the same (recorded at 295 mμ); (5) emission polarization spectrum of fluorescence (excitation 265 mμ); (12) fluorescence polarization spectrum, determining the relative contribution of the negative oscillator to absorption, at t = -196°C; (7) phosphorescence spectrum; (8) absorption polarization spectrum of phosphorescence (recorded at 460 mμ); (9) emission polarization spectrum of phosphorescence (excitation 265 mμ); (10 and 11) spontaneous emission (excitation 265 mμ) and absorption (recorded at 460 mμ) polarization spectra of phosphorescence, respectively; (6) polarization spectrum of phosphorescence determining relative contribution to absorption of the negative oscillator (recorded at 460 mμ) (Chernitskii, Konev, and Bobrovich, 1963).

appear to require some explanation. The constancy of the fluorescence spectrum of indole (and tryptophan) in water, polyvinyl alcohol, and hexane on excitation by different wavelengths in the range of 250 to 300 mμ indicates that, irrespective of the ratio of the initial populations of the ^1L_a and ^1L_b electronic levels, by the time of emission of light by the molecules some equilibrium redistribution of the energy between the electronic and vibrational levels takes place. This obviously entails the population of the ^1L_a level via the ^1L_b level and the reverse process. This alone can explain the complete similarity of the fluorescence spectra excited by light of wavelength 296 mμ (initial population of only ^1L_a level) and the fluorescence spectra excited by light of wavelength 289 mμ (initial population mainly of ^1L_b level). Although in the case of "intrinsic" absorption and "intrinsic" emission of light with the participation of only the ^1L_b level the fluorescence polarization will have the high positive values characteristic of the usual linear oscillator (40 to 50%), the result of such a redistribution of energy will be that the polarization in the region of the fluorescence of the ^1L_b transition may even become negative. In conditions in which the quantum yields of the fluorescence of ^1L_a are equal on direct excitation and excitation via the ^1L_b level, the polarization in the first case will have positive values and in the second case negative values (or zero depending on the exact size of the angle between the oscillators). Hence, the polarization of the $^1L_b \rightarrow A$ fluorescence will be determined entirely by the relative number of quanta absorbed by the ^1L_a and ^1L_b oscillators at the given wavelength. In this case, of course, the positive and negative oscillators change places. In other words, when ^1L_b fluorescence is detected in pure form, without an admixture of ^1L_a, the absorption polarization spectrum of fluorescence will be a mirror image of the usual spectrum: The maxima will be replaced by minima and vice versa. Although such an experiment cannot be conducted in pure form, owing to the mixed nature of the fluorescence at any point in the fluorescence spectrum, changes in the shape of the absorption polarization spectrum can still be expected when the fluorescence is recorded at $\lambda = 295$ mμ. Curve 4 in Fig. 13 confirms these arguments: At the former maximum of the polarization spectrum at 270 mμ there is a strong reduction of the polarization, whereas at the former minimum at 289 mμ the polarization is unaltered. It is easy to conceive that removal of the ^1L_a admixture from the fluorescence would lead to a further change in the shape of the spectrum in the same direction. The same picture is observed in the case of indole (Fig. 17, curve 4).

Thus, the excitation polarization spectra of the short-wave and long-wave parts of the fluorescence spectrum differ from one another, which is also evidence of the ability of tryptophan molecules to emit absorbed energy from two electronic levels, the transition moments of which in the ground state are differently oriented in space.

Since the ^1L_b oscillator in emission is responsible only for an insignificant fraction (5 to 10%) of the fluorescence, the main part of the excitation energy will be transferred to the ^1L_a oscillator. This is indicated by the coincidence of the tryptophan fluorescence excitation spectrum and the absorption spectrum. Such a transfer is equivalent in effect to quenching of the intrinsic ^1L_b fluorescence and, hence, to reduction of the lifetime of the corresponding excited state. If it could be shown that two excited states with considerably different lifetimes are implicated in the formation of

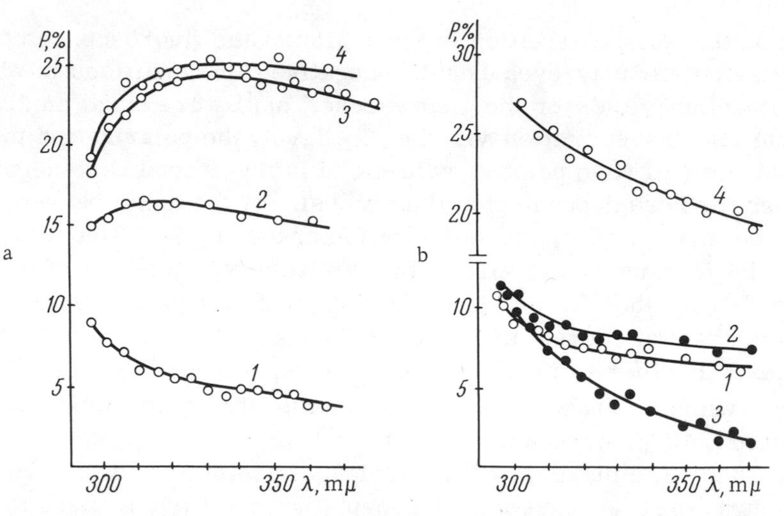

Fig. 18. Emission polarization spectra of fluorescence of: (a) a film of polyvinyl alcohol impregnated with tryptophan at concentrations of 5×10^{-1} (1), 5×10^{-2} (2), and 5×10^{-3} g/g (3), and with glycyltryptophan at a concentration of 5×10^{-3} g/g (4); (b) proteins: (1 and 2) aqueous solutions of chymotrypsinogen and trypsin, (3) wool keratin (fibers not oriented), and (4) wool keratin stretched to twice its length in water vapor and oriented parallel to the electric vector of the exciting light. In every case the fluorescence was excited at 265 mμ (Konev, Bobrovich, and Chernitskii, 1965).

the fluorescence band, it would be a strong argument in favor of the possibility of fluorescence from two electronic levels.

Thus, the question can be put in the following way: Is the τ of fluorescence constant or not over the emission spectrum?

To answer this question Konev, Bobrovich, and Chernitskii (1965) investigated the shape of the emission polarization spectrum of the fluorescence with increase in tryptophan concentration in a film of polyvinyl alcohol. Concentration depolarization is different at different points in the emission spectrum. Throughout the main part of the fluorescence band there is very efficient concentration of depolarization, except in the very short-wave region. The difference in the relationship $P_{\lambda em} = f(c)$ for different regions of the fluorescence spectrum results in the shape of the polarization spectrum's becoming opposite at high concentrations to the shape at low concentrations: The reduction in polarization in the short-wave region of the spectrum is replaced by an increase (Fig. 18a). The very weak concentration of depolarization of the fluorescence at the short-wave end of the fluorescence spectrum indicates that the lifetime of the excited ^1L_b state is much less than that for ^1L_a.

This confirms the hypothesis that two singlet excited states of tryptophan are implicated in fluorescence.

Phosphorescence of Tryptophan

Nature and Spectra of Phosphorescence. The phosphorescence of tryptophan was first discovered in 1952 by Debye and Edwards and since then has been investigated many times by a number of authors. The characteristic features of the long-wave luminescence of this amino acid indicate conclusively that it owes its origin to a triplet-singlet transition.

1. The phosphorescence spectrum is shifted in the long-wave direction compared with the fluorescence spectrum, i.e., the phosphorescence level is lower in the energy scale than the fluorescence level.

2. The phosphorescence of tryptophan has a long decay time (5.3 sec, according to Longworth), which indicates the highly forbidden nature of the triplet-singlet transition ($T^* \rightarrow S$).

3. Triplet luminescence should conform to a monomolecular exponential law of decay. This was actually observed by Debye and Edwards (1952) and by Steele and Szent-Györgyi (1957) for frozen glucose solutions of tryptophan.

4. The phosphorescence is increased by the addition of heavy atoms of potassium iodide (Vladimirov and Li Chin-kuo, 1962) and lithium bromide (Konev and Chernitskii, 1964A), which facilitate reversal of the spin by their field.

5. Tryptophan phosphorescence requires hardly any activation energy (according to the data of Steele and Szent-Györgyi (1957), E = 0.076 kcal.

6. Tryptophan phosphorescence has negative values of polarization on excitation over the whole absorption spectrum, i.e., the phosphorescence oscillator is perpendicular to the fluorescence oscillator (Chernitskii, Konev, and Bobrovich, 1963).

7. Tryptophan molecules in the phosphorescent (triplet) state have an absorption spectrum which is shifted in the long-wave direction in comparison with the ground state, as Grossweiner (1956 and 1959) showed by the method of flash photolysis.

8. Tryptophan and indole exhibit *a* phosphorescence in solution (Barenboim, 1962), in a film of polyvinyl alcohol, and in crystals. (Data obtained by the author in collaboration with E. A. Chernitskii.)

9. The concentration of unpaired electrons of tryptophan molecules in the triplet state should decrease after the light flash

in accordance with the same law as for phosphorescence. In fact, Smaller (1962) recorded photoinduced EPR (electron paramagnetic resonance) signals for indole, and Ptak and Douzou (1963), similar signals for tryptophan, which they attributed to triplet states.

According to Gribova (1964), ethanol solutions of tryptophan at 77°K exposed to the integral light of an SVDSh-1000 lamp show a tryptophan EPR signal with a decay time of about 3 sec. Before absorption of the quantum the tryptophan molecules which will later emit fluorescent or phosphorescent light do not differ in any way from one another: The phosphorescence excitation spectra coincide with the fluorescence excitation spectra (Vladimirov and Litvin, 1960).

Using a phosphoroscope with a time resolution of 5×10^{-3} sec, Freed and Salmre in 1958 were the first to obtain well-resolved phosphorescence spectra for methanol-ethanol (9:1) solutions of tryptophan, indole, indoleacetic acid, and tryptophan cooled to liquid-nitrogen temperature. These authors observed distinct tryptophan phosphorescence maxima at 408, 438, and 460 mμ. For aqueous salt solutions, maxima at 406, 432, and 456 mμ and a shoulder at 470 to 480 mμ were recorded by Vladimirov and Litvin (1960) for the long afterglow and by Vladimirov and Li Chin-kuo (1962) and Burshtein (1964) for the low-temperature luminescence (without the use of the phosphoroscope). The phosphorescence spectrum, like the fluorescence spectrum, is due entirely to the indole ring without any appreciable participation of the substituent groups. This is manifested in the similarity of the phosphorescence spectra of indole, tryptophan, and other derivatives (Freed and Salmre, 1958; and Chernitskii, Konev, and Bobrovich, 1963). Without dwelling on the details of the phosphorescence spectra, we note their main feature (Fig. 13, curve 7)—their distinct structure, in contrast to the structurelessness of the fluorescence spectra under the same conditions, despite the fact that the two spectra exhibit the same system of energy levels of the ground state and that there are no appreciable differences between the excitation spectra of these two forms (Vladimirov and Litvin, 1960). These spectral features led Vladimirov and Li Chin-kuo (1962) and Vladimirov (1964) to suggest that the phosphorescence is due to a nonequilibrium, improbable vibrational configuration of the tryptophan molecule. The whole chain of events leading to phosphorescence, according to their theory, is as follows: $S_0 \rightarrow S_1 \rightarrow T_1 \rightarrow A \rightarrow S_0$, where the $S_0 \rightarrow S_1$ transition corresponds to absorption; $S_1 \rightarrow S_0$, to fluorescence; T → S_0, to phosphorescence; and A is a nonequilibrium vibrational state. Hence, phosphorescence is due not only to spin inversion, but also to a change in configuration of the molecule. In this way the authors explain the long duration (5 sec) of tryptophan phosphorescence.

We regard this interpretation as unsatisfactory. First of all, the long duration of the phosphorescence is not due to configurational alterations, but to the probability of spin inversion by spin-orbital interaction. In the presence of heavy atoms (lithium bromide) the phosphorescence spectrum is unchanged but the decay time becomes much shorter (Konev and Chernitskii, 1964). In addition, the pronounced structure of the phosphorescence spectrum is probably not due to the appearance of an improbable configuration of the molecule but most likely reflects the ordinary state of the molecule, uncomplicated by various interactions. The reason for the noncorrespondence of the fluorescence and phosphorescence spectra becomes apparent when the question is resolved in the reverse manner. It is precisely because the fluorescence spectrum depends more on the external medium than on the internal energetic properties of the molecule that there is no similarity between the phosphorescence and fluorescence spectra. For indole in nonpolar solvents (hexane and cyclohexane) and also for tryptophan residues contained in amylase crystals at 77°K, i.e., in conditions which promote the appearance of the molecular spectrum, the position of the phosphorescence maxima in the frequency scale is not greatly altered, but the fluorescence spectrum shows frequencies characteristic of the phosphorescence spectrum. The frequency symmetry between the molecular spectrum of phosphorescence and fluorescence, on the one hand, and the spectrum of tryptophan phosphorescence in polar solvents, on the other, shows that in polar solvents the molecules in the triplet state are less energetically distorted than in the singlet state. Vladimirov's scheme is indirectly contradicted by the fact that the spectra and intensity of tryptophan and indole phosphorescence are independent of the wavelength of the exciting light in the case of monochromatic excitation in the range of 240 to 313 mμ. Hence, molecules with a different supply of vibrational energy are indistinguishable from one another in their ability to emit phosphorescence and in the spectral composition of the phosphorescence. Several facts indicate that the most important role in changes in the probabilities of triplet-singlet conversions in the tryptophan molecule is played by hydrogen of the imino group. In fact, indole in a nonpolar solvent (hexane) exhibits hardly any fluorescence and gives intense phosphorescence (free NH group). Indole and tryptophan in alcohols (methyl, ethyl, butyl, and ethylene glycol) and in aqueous salt solutions give average and equal values of relative phosphorescence intensity (an NH group which has formed a hydrogen bond), and, finally, in 0.1 M NaOH, where the imino group is completely ionized, the phosphorescence practically disappears. In all cases the shape of the excitation and phosphorescence spectra indicates that we are dealing with monomeric centers of luminescence. Hence, a gradual increase in the distance between the hydrogen and nitrogen of the imino group is accompanied by a reduction in the relative phosphorescence intensity.

The role of the imino group in determining the relative probability of singlet-triplet transitions is clearly seen by the example of tryptophan in the crystalline state. It is known that the phosphorescence of tryptophan crystals, in view of its extremely low intensity, can be observed only after it is separated from the intense fluorescence by means of a phosphoroscope (Bobrovich and Konev, 1964). Tryptophan in the crystalline state differs considerably from tryptophan in solution not only in the low intensity of the phosphorescence, but also in the fact that the phosphorescence spectrum is shifted in the long-wave direction to 500 mμ and loses its structure. Concentrated aqueous solutions of tryptophan give a similar picture. The addition of substances which bind the hydrogen of the imino group and thus prevent association (formaldehyde) leads to restoration of the ordinary phosphorescence of monomeric tryptophan. Tryptophan in the crystal lattice is apparently in the form of dimers formed with the participation of hydrogen of the imino group. This is indicated by the frequency shift of the NH vibrations in crystalline samples of tryptophan and indole to a position in the frequency scale of the infrared spectrum typical of the N · · · H—N group bound by a hydrogen bond. Moreover, judged by the excitation spectra, the absorption spectrum of crystalline tryptophan molecules is shifted 3 to 4 mμ in the long-wave direction (to 285 mμ) in comparison with monomeric forms (Bobrovich and Kembrovskii)

Alpha Phosphorescence of Tryptophan. Barenboim (1962) succeeded in recording the α phosphorescence of tryptophan nonspectrally. This emission, as is known, is due to the thermally activated return of the electron with the spin reversed from the triplet level to

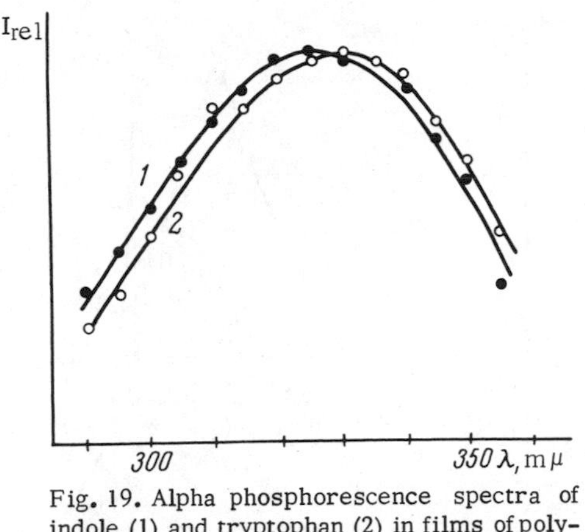

Fig. 19. Alpha phosphorescence spectra of indole (1) and tryptophan (2) in films of polyvinyl alcohol at -196°C. The resolving time of the phosphoroscope was 10^{-3} sec. (Data obtained by the author in collaboration with E. A. Chernitskii.)

the singlet level, followed by a radiative transition to the ground state. Consequently, the α phosphorescence should have the same spectrum as the fluorescence, have a long decay time, and be enhanced by temperature increase within a certain range. The last two features were actually observed in the cited work. The α phosphorescence spectrum of indole and tryptophan in a film of polyvinyl alcohol at liquid nitrogen temperature was investigated in our laboratory and was found to be similar to the fluorescence spectrum in the same conditions (Fig. 19). However, the α phosphorescence intensity obtained with a phosphoroscope with a resolving time of 10^{-2} to 10^{-3} sec was four orders weaker than that of the β phosphorescence. An intense, finely structured α phosphorescence has been recorded only for indole crystals (Chernitskii and Konev) (Fig. 20).

Absorption Polarization Spectra of Phosphorescence. In 1963 we (Chernitskii, Konev, and Bobrovich) measured the absorption polarization spectrum of tryptophan phosphorescence. The polarization spectrum obtained when the integral phosphorescence of tryptophan in a film of polyvinyl alcohol at liquid-nitrogen temperature was recorded is shown in Fig. 13 (curve 8). The fairly high negative values of the polarization of the phosphorescence over the whole absorption spectrum show that the phosphorescence oscillator makes almost a right angle with the ^1L_a oscillator and the

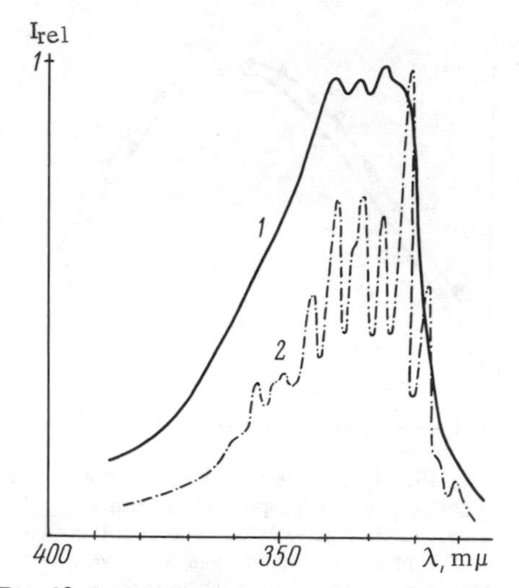

Fig. 20. Luminescence spectra of indole crystals at room temperature (1) and at t = -196°C (2). (Data obtained by the author in collaboration with E. A. Chernitskii.)

^1L_b oscillator, i.e., the phosphorescence oscillator is perpendicular to the plane of the indole ring. The oscillators corresponding to each of the three main maxima in the phosphorescence spectrum (410, 438, and 460 mμ) have a similar orientation in space, since the absorption polarization spectra of the phosphorescence recorded at these points are identical with one another. In other words, the three well-resolved phosphorescence maxima must be ascribed to the same electronic transition. The same conclusion can be derived from the constancy of τ of the phosphorescence of indole and tryptophan in ethanol for the whole phosphorescence spectrum in the range of 400 to 500 mμ (Konev, Volotovskii, and Chernitskii). For the phosphorescence of indole and tryptophan τ was 7.0 and 7.3 sec, respectively.

The small reduction of the negative values of the polarization of phosphorescence at 289 mμ indicates that the phosphorescence oscillator is not exactly perpendicular to the oscillator of the $^1L_b \rightarrow A$ electronic transition but has a small component in its direction. The slightness of this dip (1 to 2% against 6 to 7% in the fluorescence polarization spectrum) indicates that the angle between the phosphorescence oscillator and the ^1L_b absorption oscillator is almost a right angle. The existence of a negative oscillator in the long-wave absorption band is further confirmed by experimental measurements of induced polarization of the phosphorescence of anisotropic tryptophan and indole films in transmitted light. When the direction of stretching of the film coincides with that of the vibrations of the electric vector of the polarized exciting light, P_{phos} is -30%, and, when they are perpendicular to one another (rotation of the polarizer or film through 90°), P_{phos} unexpectedly takes positive values (+8%). In an isotropic film of the same concentration, P is -13% in both cases. This change of sign of the polarization indicates that molecules in an anisotropic film have absorption oscillators which are mainly perpendicular to the direction of stretching, i.e., negative ^1L_b absorption oscillators.

Anisotropic films of tryptophan-activated polyvinyl alcohol permit the use of the method which has been developed in our laboratory for the detection of the absorption spectrum of the negative ^1L_b oscillator in instances when its absorption is actively used for excitation of the phosphorescence. A spectrum representing the relationship

$$P_{\lambda fl} = f\left(\frac{\varepsilon^1 L_{b\lambda}}{\varepsilon^1 L_{a\lambda} + \varepsilon^1 L_{b\lambda}}\right)$$

is obtained by a method similar to that described in the section "Polarization Spectra of Tryptophan Fluorescence." Only in this case will the negative absorption oscillator transmit the absorbed energy to the oscillator lying across the direction of stretching, i.e., horizontally. Consequently, negatively polarized phosphorescence due to absorption by the ^1L_b oscillator of the oriented molecules is superimposed on the completely depolarized phosphorescence of the chaotically arranged molecules. However, the absorption spectrum of the negative oscillator will remain unaltered whether it is detected by polarized fluorescence or by polarized phosphorescence.

The experiment completely confirms the above arguments and again demonstrates the complex nature of the long-wave absorption band of tryptophan and indole. As curve 12 in Fig. 17 shows, the absorption polarization spectrum of phosphorescence actually lies entirely within the region of negative values of polarization and has a maximum negative value of polarization at 289 mμ. At 300 mμ, where the ^1L_b oscillator does not absorb, the polarization attains zero value.

Emission Polarization Spectra of Phosphorescence. The emission polarization spectra of the phosphorescence of tryptophan and indole, investigated in the same work as the absorption polarization spectra (Chernitskii, Konev, and Bobrovich, 1963), are shown in Figs. 13 and 17, curves 9. A common feature of these spectra is the increase in the negative values of the polarization toward the short-wave end of the phosphorescence band. At the 410 mμ maximum P_{phos} reaches -18%. At the other, the long-wave end of the phosphorescence band, the polarization values are lower, about -10%. Another characteristic feature of the polarization spectra is the presence of vibrational structure, and the polarization has maximum negative values at the phosphorescence maxima and minimum values at the minima. A polarization spectrum with such a vibrational structure has been recorded not only for tryptophan, but also for indole and indolepropionic acid. The fine structure of the phosphorescence polarization spectra becomes particularly distinct in the polarization spectra of the spontaneous polarization of the phosphorescence in anisotropic films (Figs. 13 and 17, curves 10). This structure may be due to the superposition of two vibrational series of one electronic transition, as Zimmermann et al. assumed for indole.

Oscillator Model of Tryptophan Molecule

As already mentioned, the question of the orientation of the oscillators to one another can be resolved by an analysis of the absorption and emission polarization spectra of fluorescence. The constancy of the polarization of the fluorescence of tryptophan and indole in the range of 330 to 360 mμ indicates that in this region the emission is due to one oscillator. It can be concluded from a comparison of the fluorescence polarization spectrum and the absorption spectrum of indole that this oscillator must be ^1L_a (the maximum polarization values correspond to the regions of the absorption spectrum where the contribution of the ^1L_a electronic transition is greatest). The shape of the absorption polarization spectrum of fluorescence indicates that the second oscillator of the long-wave band is ^1L_b, since it corresponds to the dip in the polarization spectrum at 289 mμ and makes an angle close to 90° with the ^1L_a oscillator (the great overlap of the ^1L_a and ^1L_b bands in the absorption spectra of indole and its derivatives in polar solvents does not permit a reliable conclusion as to whether the angle between the oscillator is 90° or less). The negative values of the polarization of the fluorescence in the region of the second absorption band at 220 mμ and the negative sign of its dichroism indicate that the oscillator of this band is perpendicular to the ^1L_a oscillator. The negative values of the polarization of the phosphorescence on excitation at any point of the absorption spectrum indicate that the phosphorescence oscillator is perpendicular to the absorption and fluorescence oscillators.

Some evidence in support of the expressed hypotheses regarding the mutual orientation of the absorption and emission oscillators is given by experiments with oriented molecules of tryptophan, indole, and indolepropionic acid. The orientation of the molecules was effected by mechanical stretching of polyvinyl alcohol films impregnated with indole or tryptophan. The intention was to find out how the fluorescence oscillator is oriented to the direction of stretching of the film. For this purpose the degree of spontaneous polarization of the fluorescence was measured in relation to the film azimuth, i.e., the angle between the vertical and the direction of stretching of the film. The relationship $P_{fl} = f(\varphi)$ was recorded for tryptophan (Fig. 21), indole, and acetyltryptophan. The figure shows that maximum positive values were observed at $\varphi = 0°$ and maximum negative values at $\varphi = 90°$. At $\varphi = 45°$ the polarization was zero. The same relationships were found for acetyltryptophan and indole. Since in these experiments the fluorescence was recorded in the range of 330 to 360 mμ, where the polarization is constant, and since the maximum positive values of this fluorescence occurred where the relative contribution of the ^1L_a oscillator was greatest

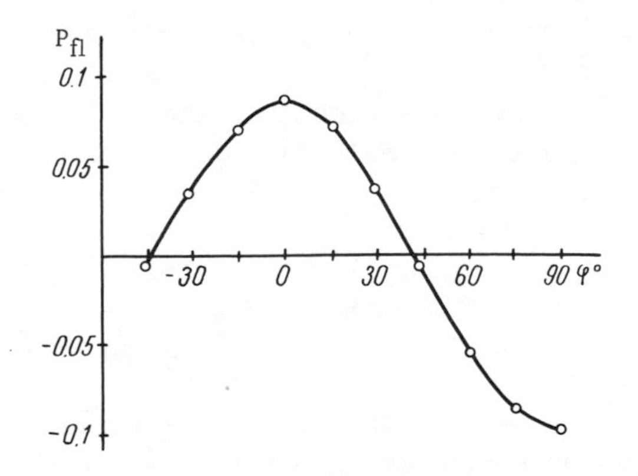

Fig. 21. P_{fl} of tryptophan fluorescence as a function of azimuth of film, t = 20°C (Chernitskii and Konev, 1965).

(265 and 300 mμ), firstly, we are clearly dealing with the ^1L_a oscillator in fluorescence, and, secondly, this oscillator is oriented along the direction of stretching of the film. The maximum values of the dichroic ratio of these films occurred at $\varphi = 0°$. This corresponds to the general rule for all organic molecules, where the long-wave absorption oscillator coincides with the fluorescence oscillator.

Thus, tryptophan and indole molecules are oriented in the film in such a way that the ^1L_a oscillator in absorption and emission lies along the direction of stretching. Another thing to be determined is the way in which the tryptophan molecules themselves are oriented in the film to attach this oscillator to the molecular skeleton. The most obvious explanation is that the stream of polymer threads sliding past one another orient the tryptophan residues with the long axis along the direction of stretching. In this case substituents which lengthen the molecule should facilitate orientation in the film. However, the opposite was found in the experiment. The spontaneous polarization of the fluorescence of films stretched to the same extent (fivefold stretching) was higher for indole (20%) than for its derivatives (only 13% for tryptophan, for instance). Indole-impregnated films had higher values for spontaneous polarization of the fluorescence and dichroism. Hence, indole and tryptophan molecules are oriented with their short axis along the direction of stretching, and the side chain of tryptophan impedes orientation. This hypothesis is confirmed by the ability of the hydrogen of the imino group to form hydrogen bonds. This will inevitably obstruct orientation with the long axis of the molecule in the direction of stretching and, conversely, assist orientation of the short axis.

Thus, the ^1L_a oscillator lies along the short axis of the molecule. According to the above, this oscillator is also oriented along the direction of stretching of the film. On the other hand, this conclusion agrees well with the previous conclusions derived from the different shift of the ^1L_a and ^1L_b absorption bands on changeover from hexane to alcohol solutions: The ^1L_b oscillator is directed along the long axis of the molecule, and the ^1L_a oscillator, along the short axis.

It should be noted that the indicated position of the oscillator found from the experimental data is in complete agreement with theoretical predictions based on a simplified model of the perimeter. It was assumed for naphthalene that the ^1L_a transition is oriented along the short axis, and ^1L_b, along the long axis (Platt, 1949). Assuming that indole and naphthalene are iso-π-electron systems and their spectral characteristics are similar, Platt (1951) assumed that for them the orientation of the electronic transition oscillators is identical. Platt's viewpoint was subsequently accepted by Zimmermann and Joop (1961). Weber, in Shifrin's opinion (1964), holds similar views.

However, in indole and tryptophan the ^1L_b oscillator does not lie exactly along the short axis of the molecule, but at an acute angle to it; the angle between the ^1L_a and ^1L_b oscillators is less than 90°. This can be inferred from the shape of the absorption polarization spectra of the fluorescence for indole and its derivatives. The absorption spectrum of indole in hexane indicates that the molar extinctions of the ^1L_a and ^1L_b bands are approximately the same. According to the shape of the absorption spectra in alcohol solutions and solid films of polyvinyl alcohol, the contribution of the ^1L_b oscillator to absorption is not less than that of ^1L_a in the 289-mμ region. If they are mutually perpendicular, this will be accompanied by complete depolarization of the fluorescence at this point, but this is not actually the case. Moreover, since the ^1L_a and ^1L_b absorption oscillators lie in the plane of the molecule and the ^1L_a oscillator is oriented along the direction of stretching of the film, then, if the oscillators are mutually perpendicular, the absorption spectra should be different when they are measured in horizontally and vertically polarized light. The shape of the dichroism spectrum

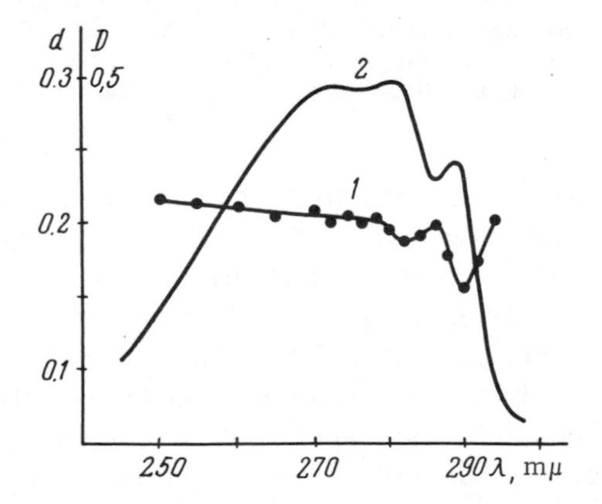

Fig. 22. Dichroism of anisotropic film of indole-impregnated polyvinyl alcohol film (1) and its absorption spectrum (2) (Chernitskii and Konev, 1965).

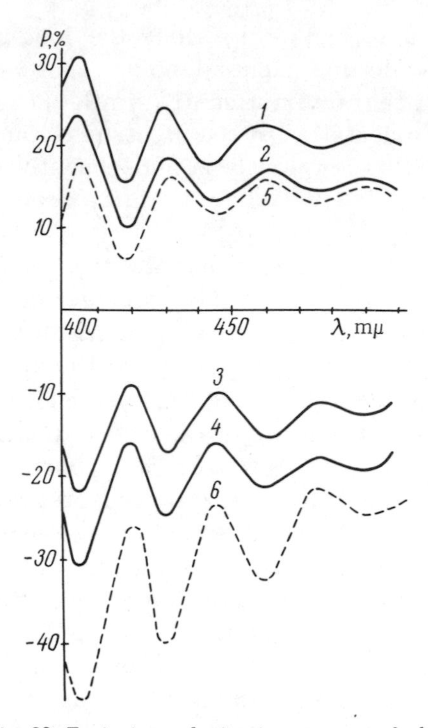

Fig. 23. Emission polarization spectra of phosphorescence of indole-activated anisotropic films of polyvinyl alcohol ("transillumination" measurements): (6) Direction of stretching of the film coincides with that of the electric vector of the exciting light; (5) film turned through 90° in its plane and spectra of spontaneous polarization of phosphorescence for these films in relation to the azimuth of the film: (1) $\varphi = 90°$, (2) $\varphi = 70°$, (3) $\varphi = 20°$, (4) $\varphi = 0°$. (Data obtained by the author in collaboration with E. A. Chernitskii.)

in this case will correspond to that of the polarization spectrum, since this is observed with most dyes (Feofilov, 1961).

In fact, for indole there are two minima, at 282 and 290 mμ, in the dichroism spectrum (Fig. 22). However, the absolute value of these minima is much smaller than in the fluorescence polarization spectrum. The failure of the fluorescence polarization spectrum to coincide with the dichroism spectrum becomes understandable if the ^1L_b oscillator is assumed to be oriented at an angle of a little less than 90° to the direction of the ^1L_a oscillator.

Thus, some data indicate that the angle between the ^1L_a and

^1L_b oscillators for indole and its derivatives is less than 90° and lies between 45 and 90°.

After allocation of the ^1L_b oscillator to the long-wave axis of the indole molecule, the allocation of the ^3L_a oscillator, corresponding to a transition from a metastable to the ground state, can be made only on the basis of experimental investigations of the polarization spectra of isotropic films. It can be concluded from the shape of the fluorescence polarization spectrum obtained in our laboratory (Fig. 17, curve 8) that the phosphorescence oscillator is mainly perpendicular to the ^1L_a and ^1L_b absorption oscillators. Anisotropic films provide further confirmation of this. The perpendicularity of the phosphorescence oscillator to the ^1L_a oscillator is shown by experimental investigations of the spontaneous polarization in relation to the azimuth of the film. The maximum values of polarization at $\varphi = 90°$ conclusively demonstrate that the phosphorescence oscillator is perpendicular to the ^1L_a oscillator (Fig. 23).

The small dip in the 289-mμ region and the bulge at 296 mμ in the fluorescence polarization spectra of isotropic films indicates that the phosphorescence oscillator is not exactly perpendicular to the other oscillator, the ^1L_b oscillator. This conclusion is also confirmed by the fact that, when phosphorescence is excited by the 265-mμ mercury line, the emission polarization spectrum of the phosphorescence of indole is almost a straight line, whereas excitation by the 280-mμ line produces a spectrum with a distinct structure, which is presumably due to the superposition of two series of electronic vibrational bands polarized in mutually perpendicular directions. The 280-mμ line, of course, excites a relatively larger proportion of ^1L_b oscillators which have a phosphorescence component coinciding in direction with them. This leads to a reduction in the polarization at the corresponding points of the spectrum. When the ^1L_a oscillator (only or predominantly) is excited, the oscillators of the two vibrational series of the phosphorescence band are oriented perpendicular to it, which leads to constancy of the negative values of the polarization over the whole phosphorescence spectrum.

Similar relations are observed in the experiment shown in Fig. 24.

Finally, we shall briefly discuss the reasons for identifying the phosphorescent state of tryptophan as ^3L_a. It is known that, on the basis of knowledge of the selection rules for the T-S transition partially allowed by spin-orbital interaction and the values of the

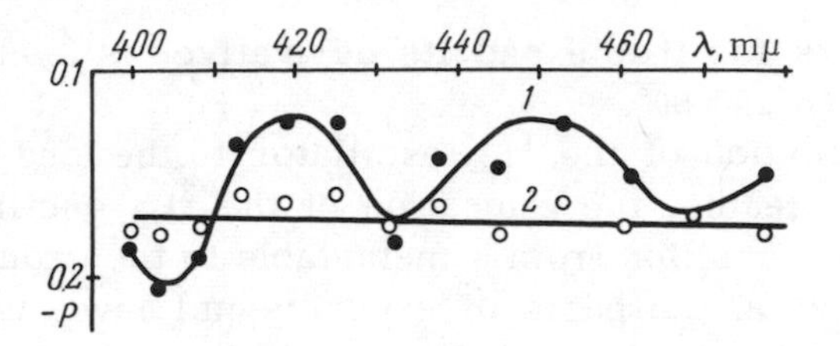

Fig. 24. Emission polarization spectra of phosphorescence of indole in films of polyvinyl alcohol excited by λ = 280 mμ (1) and λ = 265 mμ (2) (Chernitskii and Konev, 1965).

polarization of the phosphorescence for excitation of ^1L_a and ^1L_b at the maxima, the symmetry of the phosphorescent state can be established (Williams, 1959). Theoretical calculations indicate that the lowest triplet state of aromatic hydrocarbons must be ^3L_a (Pariser, 1956). Experiments provide good confirmation of the theory for such a compound as phenanthrene (Azumi and McClynn, 1962).

In the case of tryptophan and indole additional confirmation of the nature of the T-S transition is the partial correspondence of the phosphorescence and ^1L_a fluorescence of amylase and indole crystals in cyclohexane.

We can sum up the foregoing as follows (Fig. 25): The ^1L_b oscillator, which corresponds to the $^1L_b \leftarrow A$ electronic transition with a maximum at 289 mμ in the long-wave absorption band and to the $^1L_b \rightarrow A$ electronic transition in the short-wave part of the fluorescence band, is oriented in the plane of the molecule along its long axis. The ^1L_a oscillator, which corresponds to the $^1L_a \leftarrow A$

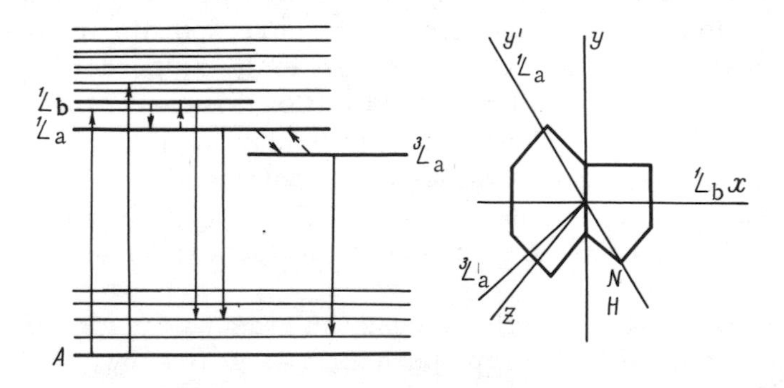

Fig. 25. Scheme of electronic vibrational levels of indole and tryptophan molecules and orientation of corresponding oscillators to the "skeleton" of the molecule.

electronic transition with an absorption maximum at 266 mμ (in hexane) and a maximum at 272 mμ in polar solvents, and to the $^1L_a \rightarrow A$ electronic transition in emission (it is entirely responsible for the fluorescence in aqueous solutions), lies in the plane of the molecule at an angle of less than 90° relative to the ^1L_b oscillator. There is efficient redistribution of energy between the ^1L_a and ^1L_b electronic levels, which are jointly implicated in fluorescence. The ^1B_b oscillator, corresponding to the $^1L_b \leftarrow A$ transition with a maximum at 220 mμ in absorption, lies at an angle of almost 90° to the ^1L_a oscillator. The ^3L_a phosphorescence oscillator is perpendicular to the plane of the molecule and has a small component along the long molecular axis, i.e., it is perpendicular to the ^1L_a oscillator but is not exactly perpendicular to the ^1L_b oscillator. The unusual conclusion which emerges from an investigation of the luminescent properties of tryptophan is that the fluorescence may be due to two electronic levels—that there are two singlet electronic excited states of this amino acid. As is known, most organic compounds are characterized by radiative transitions from the zero vibrational sublevel of the lower electronic level, irrespective of the site of initial excitation in the energy scale. Ideas regarding the possibility of the emission of fluorescence quanta from two levels for indole molecules are now being developed (Zimmermann and Geisenfelder, 1961; Schütt and Zimmermann, 1963; and Zimmermann and Joop, 1961).

It would probably be incorrect to regard indole and its derivatives as the only compounds capable of fluorescing from two levels. Fluorescence from a second level has been found in the case of azulene (Zimmermann and Joop, 1960) and from a first and second level in the case of paradisubstituted benzene derivatives (Lippert, Luder, and Boos, 1962; and Roe, 1961), and p-cyano-dimethylaniline (Lippert, 1962).

TYROSINE

Many authors have investigated the fluorescence of tyrosine (Fig. 26) in aqueous solutions (Bowman et al., 1955; Vladimirov and Konev, 1957; Cowgill, 1963; and Teale and Weber, 1957) and in crystals (Brumberg, 1956; and Vladimirov, 1961). The maximum of the fluorescence band of aqueous solutions is at 303 mμ. When aqueous solutions are frozen, the fluorescence maximum is shifted in the short-wave direction and lies at 298 mμ (Vladimirov and Li Chin-kuo, 1962). Tyrosine crystals exhibit the same fluorescence spectrum as an aqueous solution (Barskii and Brumberg, 1958). Lehrer and Fasman (1964) reported excimer fluorescence of polytyrosine at 420 mμ.

The fluorescence excitation spectrum coincides with the absorption spectrum, i.e., maxima at 275 and 222 mμ (Teale and Weber, 1957). According to the data of these authors, the quantum yield for any exciting wavelength is 0.21.

Cowgill (1963), who apparently did not introduce any correction for the spectral response of the apparatus, nevertheless indicated a correspondence between the position of the fluorescence maxima of tyrosine and that of m-cresol, OH-benzyl alcohol, phenol, L-tyrosylglycine, glycyl-L-tyrosine, tyramine, glycyl-L-tyrosylglycinamide, and leucyl-L-tyrosine (320 mμ). Hence, the position of the fluorescence maximum is determined exclusively by the phenol nucleus without any effect from the substituents. The situation is different with the fluorescence quantum yield. The substituent groups can have an appreciable effect on the absolute

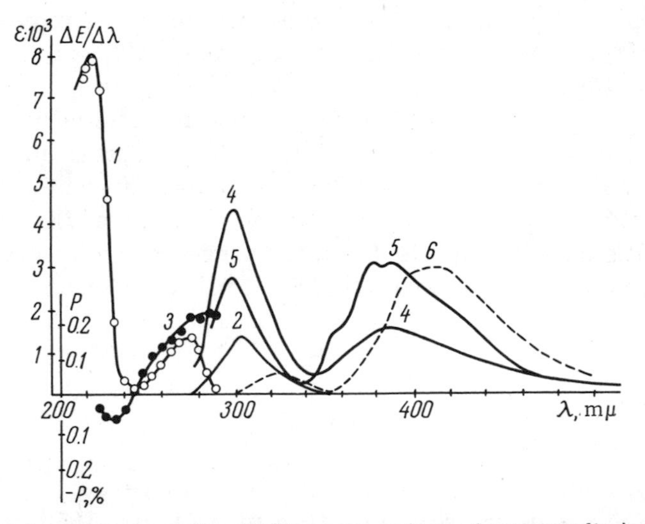

Fig. 26. Tyrosine, aqueous solution: (1) Absorption spectrum (continuous line) and fluorescence excitation spectrum (dashes); (2) fluorescence spectrum (Teale and Weber, 1957); (3) absorption polarization spectrum of fluorescence (propylene glycol, excitation by nonpolarized light; Weber, 1960); (4) luminescence spectrum at t = -140°C (Vladimirov and Burshtein, 1960); (5 and 6) luminescence spectra in 0.1 M glycocoll and 0.1 N NaOH, respectively, at t = 77°K (Vladimirov and Li Chin-kuo, 1962).

Table 3. Effect of Substituents on Quantum Yield of Fluorescence of Tyrosine Derivatives (Cowgill, 1963)

$$HO-C_6H_4-CH_2-CH(R_2)-R_1$$

Compound	R_1	R_2	φ
Tyrosine	$-COO^-$	$-\overset{+}{N}H_3$	0.21
Tyramine	$-H$	$-\overset{+}{N}H_3$	0.185
Leucyltyrosine	$-COO^-$	$_3H\overset{+}{N}-CH(leuc)-CO-NH-$	0.103
Tyrosylglycine	$-CO-NH-CH_2-COO^-$	$-\overset{+}{N}H_3$	0.074
Glycyltyrosine	$-COO^-$	$_3H\overset{+}{N}-CH_2-CO-NH-$	0.070
Tyrosine	$-COOH$	$-\overset{+}{N}H_3$	0.056
Glycyltyrosyl-glycinamide	$-CO-NH-CH_2-CO-NH_2$	$_3H\overset{+}{N}-CH_2-CO-NH-$	0.035
Glycyltyrosine	$-COOH$	$_3H\overset{+}{N}-CH_2-CO-NH-$	0.027

yield. The un-ionized carboxyl group quenches the fluorescence very strongly (Table 3).

The presence of dissociable ionogenic groups in the tyrosine molecule accounts for the dependence of the fluorescence quantum yield on the pH of the medium (Fig. 9). The relationship between the fluorescence quantum yield and pH has been investigated by White (1959), Vladimirov and Li Chin-kuo (1962), and Cowgill (1963). The reduction of quantum yield in the region of low pH, beginning at 4.0, is due to conversion of the carboxyl group from the ionized state to the un-ionized state, $R-COO^- \rightarrow R-COOH$.

Tyrosine with un-ionized carboxyl groups has a quantum yield of 0.056 (Cowgill, 1963). The pF of this transition, measured from the luminescence intensity, agrees with the pK measured by amperometric titration—2.2, according to Cowgill, and 2.45, according to White.

The region of the second reduction in the quantum yield of tyrosine fluorescence is connected with the dissociation of the phenol hydroxyl and has a pK = 9.7, according to White. White, Cowgill, Vladimirov, and Li Chin-kuo believe that the dissociated form of tyrosine is incapable of luminescence.

However, Cornog and Adams (1963) observed weak fluorescence of tyrosine in 0.12 N NaOH with a maximum at 345 ± 5 mμ and a quantum yield of about 0.01 and attributed it to the phenol ion of tyrosine. This corresponds to the earlier data of Longin (1959) on the fluorescence of alkaline tyrosine in the 400-mμ region. Hence, in addition to quenching of the fluorescence of tyrosine in the un-ionized form, there is weak fluorescence of the ion

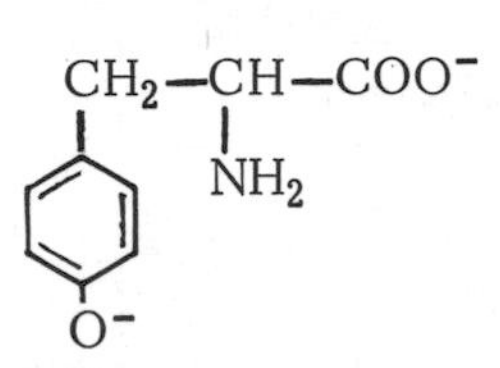

The maximum of the low-temperature fluorescence (77°K) of tyrosine in 0.1 N NaOH at 320 mμ and the phosphorescence maximum at 410 mμ probably correspond to the latter form (Vladimirov and Li Chin-kuo, 1962).

Weber (1960) investigated the fluorescence polarization spectra of tyrosine. According to his data, tyrosine and cresol in propylene glycol at -70°C have identical polarization spectra (Fig. 26), which indicate that the long-wave absorption band is formed by one electronic transition. The absolute values of the polarization within the 275-mμ absorption band, 0.21 converted to linearly polarized excitation by the formula $P = 2P_u/(P_u + 1)$, require values of about 35%, i.e., values characteristic of asymmetric molecules and molecules with second-order symmetry. It is known that a plane molecule with an axis of symmetry of the third order or higher cannot have a maximum fluorescence polarization of more than 14.3%, i.e., the value obtained by calculation for the classical model of a rotating linear oscillator and a quantum-mechanical calculation of the luminescence of diatomic molecules (Feofilov, 1961). In fact, the planar benzene molecule, which has an axis of symmetry of the sixth order perpendicular to the plane of the molecule (symmetry D_{6h}), according to Feofilov, has low values of polarization (7.7%). Phenol also has low values of polarization. Hence, only the attachment of the alanine residue in the para position is accompanied by pronounced changes in the symmetry of the electron cloud of the benzene ring on transition from a planar to a linear oscillator. It is characteristic that there is no shift of levels in the energy scale in this case. The oscillator corresponding to the second electronic transition in the 222-mμ band is oriented at a right angle to the long-wave absorption oscillator, since Weber observed negative polarizations in the 230-mμ region.

The phosphorescence of tyrosine was first observed by Debye and Edwards (1952). The tyrosine phosphorescence spectrum recorded by Steele and Szent-Györgyi (1958) had a broad maximum in the 370- to 410-mμ region. Vladimirov and Litvin (1960) and Vladimirov and Burshtein (1960) observed a maximum at 387 mμ and a weak shoulder at 410 to 420 mμ in the phosphorescence spectrum of tyrosine at -100°C.

The data of the Soviet authors were confirmed by Longworth (1961), who obtained the position of the phosphorescence maximum at 388 mμ. Under certain conditions the structureless phosphorescence band of tyrosine can be converted into a spectrum with well-resolved vibrational maxima. Some elements of the phosphorescence structure appear in a strongly alkaline medium (Vladimirov and Li Chin-kuo, 1962), and in 8 M lithium iodide, which facilitates singlet-triplet conversion, sharp phosphorescence maxima are found at 354, 366, 376, 387, 397, and 412 mμ (Konev and Bobrovich).

The decay time τ of the phosphorescence of tyrosine in neutral and acid media is 3 sec, and in an alkaline medium, 0.9 sec (Debye and Edwards, 1952). More careful measurements of τ of the phosphorescence by Longworth (1961) gave 2.1 ± 0.1 sec. It is characteristic that in a strongly alkaline medium the relative

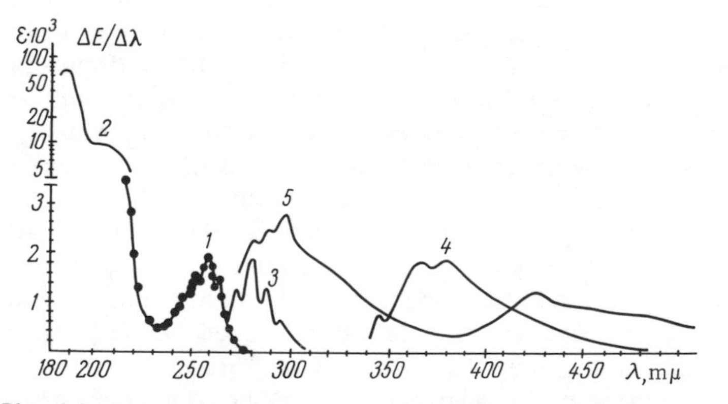

Fig. 27. Phenylalanine, aqueous solution: (1) absorption spectrum (continuous line) and fluorescence excitation spectrum (points) (Teale and Weber, 1957); (2) absorption spectrum (Wetlaufer, 1962); (3) fluorescence spectrum in a mixture of acetic acid and ethanol (1:4); (4) phosphorescence spectrum in 0.1 M glycocoll at 77°K (Vladimirov and Li Chin-kuo, 1962); and (5) luminescence spectrum of aqueous solution at t = -140°C (Burshtein, 1964).

proportion of phosphorescence in the total luminescence increases from 50 to 80% (Vladimirov and Li Chin-kuo, 1962).

The shortening of the lifetime of triplet states and the relative enhancement of phosphorescence may be due to the greater probability of singlet-triplet transitions when the phenol group is ionized.

PHENYLALANINE

Phenylalanine in aqueous solutions has a long-wave absorption band with the main maximum at 258 mμ (Fig. 27). The absorption band is distinctly structured and in the 230- to 280-mμ interval is resolved into eight elements of vibrational structure. However, the low molar extinction values of around 200 indicate that the electronic transition responsible for the long-wave absorption band is partially forbidden.

Like the absorption spectrum, the fluorescence spectrum of phenylalanine is distinctly structured. In contrast to Teale and Weber (1957), who could not manage to resolve the structure of the fluorescence spectrum, Vladimirov (1959) and Vladimirov and Burshtein (1960) recorded maxima of the vibrational structure at 282, 285, and 289 mμ and a shoulder at 303 to 305 mμ.

The ratio of the intensities of the individual elements of the fine structure of the fluorescence spectrum of phenylalanine depends strongly on the properties of the solvent: In a mixture of acetic acid and ethyl alcohol (1:4) at 77°K, the main maximum is at 282 mμ and subsidiary maxima lie at 273, 288, and 295 mμ, whereas in 0.1 N NaOH the spectrum is shifted in the long-wave direction and is formed by the main maximum at 292 mμ and a maximum at 303 mμ (Vladimirov and Li Chin-kuo, 1962).

The quantum yield of the fluorescence of phenylalanine is low, about 0.04, according to Teale and Weber (1957), and is constant over the whole excitation spectrum. In highly alkaline and highly acid media the fluorescence is slightly quenched (Vladimirov and Li Chin-kuo, 1962; and Feitelson, 1964) (Fig. 9).

Replacement of the hydrogen of the benzene ring by halogens (fluorine) leads to a marked increase in the quantum yield—to 0.4 (Guroff, Michael, and Chirigos, 1962).

The decay time of phenylalanine fluorescence has not been

determined by direct methods, but, according to the integral of the absorption band, it is $8 \cdot 10^{-9}$ sec (Beaven, 1961).

In the crystalline state the fluorescence band of phenylalanine is shifted through 15 mμ in the long-wave direction (Barskii and Brumberg, 1958; and Vladimirov, 1959), although in this case the corrections introduced for the spectral sensitivity of the apparatus cannot be considered as reliable.

The similarity of the fluorescence and absorption spectra of phenylalanine indicates that the two bands are formed by one electronic transition. The phosphorescence spectrum of phenylalanine consists of three maxima: 424 to 426, 448 to 450, and 490 to 495 mμ (Vladimirov and Burshtein, 1960). Although the phosphorescence decay time has not been determined exactly, it is probably very small, less than 0.1 sec, according to the data of Debye and Edwards (1952) and Steele and Szent-Györgyi (1957). Taking into account the fact that, according to Sveshnikov's data (1951), both benzene and phenol have long decay times τ — 4.1 and 2.2 sec, respectively — we can surmise that the alanine part of phenylalanine is capable of removing the quantum-mechanical prohibitions of the triplet-singlet transition, as halogens do, for instance (chlorobenzene has $\tau = 0.005$ and tribromobenzene has $\tau = 0.0007$ sec).

The benzene molecule shows not only intercombination prohibitions, but also symmetry prohibitions, and the latter lead to the absence of 0-0 transitions in the fluorescence and phosphorescence spectra (Sveshnikov, 1951).

It is quite possible, however, that the decay time of phenylalanine phosphorescence was not correctly determined by the cited authors, since Longworth (1961) obtained a value of 7.3 sec and Nag-Chaudhuri and Augenstein (1964) obtained a value of 5.5 ± 0.48 sec.

The polarization spectra and the orientation of the electronic transitions of phenylalanine have not been investigated.

Chapter 2

ELECTRONIC EXCITED STATES OF PROTEINS

The protein macromolecule contains a large number of chromophoric groupings, i.e., energetically independent assemblies of atoms capable of absorbing light quanta in the ultraviolet region of the spectrum. All the chromophoric groupings of proteins can be divided into two classes: (1) monomer chromophores, i.e., the 20 amino acid residues, which are capable of interacting with light both in the free state and as part of a polymer; (2) chromophores which are characteristic of the polymer itself and which disappear completely when the polymer molecule is broken up into its constituent monomers. Chromophores of the second class include atomic groupings, such as the peptide bond and the thioester linkage of mercaptides of the cysteine or glutathione type

$$\begin{array}{c} R_1{-}S{-}\underset{\displaystyle R_2}{\underset{|}{C}}{=}CH{-}R_3 \end{array}$$

Table 4 sums up some of the characteristics of the absorption properties of the main chromophoric groupings of the protein molecule (the data given refer to aqueous solutions).

Acyclic amino acids exhibit weak absorption beginning at 230 mμ.

In most proteins the majority of quanta in the 250- to 300-mμ region are absorbed by aromatic amino acids: tyrosine, tryptophan, and phenylalanine. Sulfur-containing amino acids absorb in the 235- to 250 mμ region. Absorption of light by peptide bonds begins to predominate only at the boundary of the vacuum ultra-

Table 4. Chromophoric Groupings of Proteins

Chromophore	Absorption maxima, mμ	Molar extinction at maxima	Long-wave limit of absorption spectrum, mμ	Reference
Tryptophan (pH 2)	280, 218	6500, 27,000	310	Beaven, 1961
Tyrosine (pH 2)	275, 222	1290, 8000	295	Beaven, 1961
Tyrosine (pH 12)	295, 240	2300, 10,000	330	
Tyrosine (aqueous)	275	1520 (1340)	295	Cowgill, 1963; Teale, 1962
Phenylalanine	258, 205	200, 8500	270	Beaven, 1961
Histidine	240			Edsall et al., 1954
Cysteine (alkaline)	235	4000–6000	250	Benesch and Benesch 1955; Gorin, 1956
Cystine	245–249	340–360	300	Fromageot and Schenk, 1950
Peptide bond	185	2700, 6500	220	Preiss and Setlow, 1956; Saidel, 1955; Ham and Platt, 1952
Thioester bond	308	7000	–	Brady, 1963
Acyclic amino acids	–	–	230	Magill et al., 1937
Glycine	–	10 at 220 mμ	230	Magill et al., 1937
Leucine	–	50 at 220 mμ	230	Magill et al., 1937

violet at 150 to 220 mμ. While aromatic amino acids play the leading role in light absorption, their role is even more important in fluorescence. As mentioned in the Introduction, there are no grounds for the hypotheses that all amino acids are capable of luminescence and that the protein molecule as a whole, like a semiconducting crystal, is capable of luminescence (Jordan, 1938; and Szent-Györgyi, 1941).

In 1957 I showed that the ultraviolet luminescence of proteins is due entirely to aromatic amino acids. Investigations of several plant and animal proteins—human serum albumin, egg albumin, gamma globulin from pumpkin seeds, gamma globulin of rabbit-blood serum, arachin, zein, gliadin, casein, sturin, calf-thymus histone, clupein, pepsin, and bromelin—confirmed that ability to luminesce is a common property of the vast majority of proteinaceous substances and is due to the aromatic amino acids, primarily tryptophan, contained in them.

It was found that proteins containing no aromatic amino acids—clupein and sturin—were incapable of luminescence. This immediately ruled out the possibility of luminescence of peptide bonds, acyclic amino acids, and other components of protein, as well as the possibility of luminescence characteristic of the macromolecule as a whole in accordance, for instance, with Jordan and with Szent-Györgyi's hypothesis (1941, 1955) of its semiconducting nature.

Proteins contaning tyrosine and lacking tryptophan (nucleohistone and zein) gave a tyrosine fluorescence spectrum. Tryptophan-containing proteins—pepsin, arachin, ovalbumin, etc.—exhibited a predominantly tryptophan fluorescence spectrum.

The second piece of evidence in support of the aromatic nature of the luminescence of proteins was provided by their fluorescence excitation spectra, which showed a distinct tryptophan absorption maximum at 280 mμ. This fundamental conclusion regarding the aromatic nature of protein luminescence has subsequently been confirmed several times by various scalar and vectorial characteristics of the luminescence: the spectra of the phosphorescence and its excitation spectra (Vladimirov and Litvin, 1960); the absorption polarization spectra of the fluorescence (Teale, 1960; Weber, 1960; and Konev and Katibnikov, 1961); the emission polarization spectra of the fluorescence (Konev, Bobrovich, and Chernitskii, 1964); and the emission polarization spectra of the phosphorescence (Konev and Bobrovich, 1964). For convenience of exposition and in view of the pronounced differences we follow Weber (1960) and divide all proteins into two large classes, A and B. Class A includes all proteins which contain no tryptophan residues but have the spectral luminescent properties which are due to tyrosine. Class B includes tryptophan-containing proteins; their luminescence is due mainly to residues of this aromatic amino acid alone.

CLASS A PROTEINS (Tyrosine-Containing Proteins)

The fluorescence of proteins which contain no tryptophan residues but contain phenylalanine and tyrosine is due entirely to the tyrosine residues (Konev, 1957). Weber (1960, 1961) assigned proteins of this kind to class A. Investigated proteins of this class include zein, insulin, tropomyosin, ribonuclease, trypsin inhibitor, 5,3-ketosteroid isomerase, ovomucoid, nucleohistones, malate dehydrogenase, and oxytocin (a polypeptide). According to Teale's measurements (1960), insulin, zein, ribonuclease, and ovomucoid have a tyrosine fluorescence maximum at 304 mμ. In class A proteins, as in the dipeptide phenylalaninetyrosine, the phenylalanine component is absent in the fluorescence and excitation spectra (Cowgill, 1963).

In contrast to tryptophan, the position of the tyrosine fluorescence spectrum is not affected by conformational changes of the macromolecule. The 304-mμ maximum is observed in all proteins both in the native state and after denaturation by 8 M urea. The incorporation of tyrosine in protein is accompanied only by a fluorescence-quenching effect. The quantum yield falls to 8.0, 3.7, 1.75, 1.2, and 0.7% in zein, insulin, ribonuclease, ovomucoid, and trypsin inhibitor, respectively. It might be assumed that reduction of the quantum yield is due to quenching processes in the course of tyrosine-tyrosine energy migration. In fact, the absorption polarization spectra of the fluorescence of tyrosine, which have the same shape as the corresponding spectrum of tyrosine in the free state, differ from it in the lower values of the polarization (Weber, 1960). For instance, in the case of insulin the polarization is 0.18, whereas for free tyrosine it is 0.36. Low polarization values are still found in proteins in a propylene glycol-water mixture at -70°C, i.e., under conditions which rule out the possibility of rotational depolarization. Hence, the depolarizing effect is due to tyrosine-tyrosine energy migration. Its efficiency, assessed by a method which has been described more fully for tryptophan-tryptophan energy migration, is ~ 50%. However, energy migration processes are not the cause of the low values of the quantum yields of class A proteins. This is indicated by the discovery in our laboratory that there is no concentration quenching of tyrosine fluorescence in solid films of polyvinyl alcohol, and the quantum yield of the fluorescence of the dipeptide tyrosyltyrosine (Cowgill, 1963A), where the tyrosine residues are in direct contact, is close to that of other tyrosine dipeptides (Table 5).

One of the main reasons for the low quantum yield of tyrosine residues in protein is the effect of neighboring electronegative groupings.

As long ago as 1957 I observed that the quantum yield of the fluorescence of the dipeptide glycyltyrosine was approximately half of that of free tyrosine, and I attributed the reduction of the quantum yield to the quenching effect of the nearest peptide bond.

This viewpoint was later substantiated by the systematic investigations of Cowgill (1964).

The presence of three peptide bonds causes a reduction of the quantum yield from 0.21 in pure tyrosine to 0.035 in glycyltyrosylglycinamide (Table 5). The latter quantum yield value is very close to that of class A proteins.

It would be incorrect, however, to attribute the entire quenching of the tyrosine fluorescence of proteins solely to the effect of the nearest peptide bond. From this viewpoint, why, e.g., the quantum yields of tyrosine in different proteins differ considerably from one another would be incomprehensible. The second main reason for quenching of the fluorescence of tyrosine residues in proteins is the involvement of the phenol group in a hydrogen bond, followed by the ionization of this group in the excited state and the formation of a weakly luminescent phenolate ion:

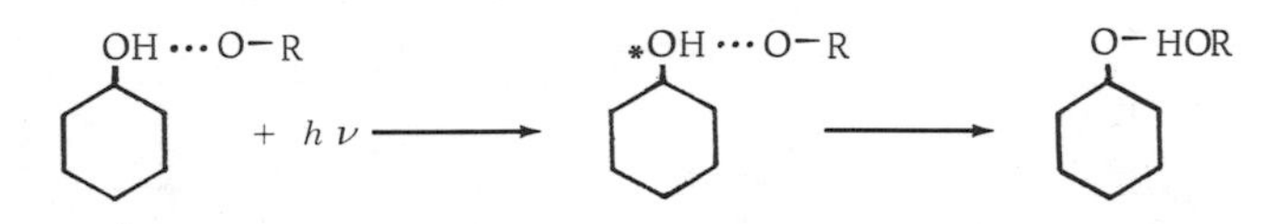

A hydrogen bond is usually formed between the OH group and the ionized carboxyl group of glutamic and aspartic amino acids.

Several facts indicate the correctness of such a point of view. Firstly, the fluorescence of tyrosine and tyrosine-containing proteins is selectively quenched by substances containing charged carboxyl groups (Teale, 1961).

Secondly, the quantum yield of tyrosine fluorescence in a copolymer consisting of 4% tyrosine residues and 96% glutamic acid residues is very low in a neutral solution (0.02) and increases at pH = 2.0 to 0.3. Hence, the conversion of the COO^- group to the COOH form, which cannot form a hydrogen bond with the OH group of tyrosine, leads to removal of the quenching effect (Rosenheck and Weber, 1961).

Thirdly, the degree of quenching of tyrosine fluorescence in different proteins is correlated with the number of carboxyl groups of aminodicarboxylic acids (Teale, 1961) per tyrosine residue (Table 6).

Fourthly, the breakage of hydrogen bonds in proteins by treatment with a mixture of alcohol and hydrochloric acid leads to an

Table 5. Quantum Yields of Fluorescence of Some Tyrosine-Containing Peptides

$$HO-C_6H_4-CH_2-CH(R_2)-R_1$$

Compound	R_1	R_2	Quantum yield	Reference
Tyrosyl	$-COO^-$	$-\overset{+}{N}H_3$	0.21	Cowgill, 1963
Tyramine	$-H$	$-\overset{+}{N}H_3$	0.185	Cowgill, 1963
Leucyltyrosine	$-COO^-$	$_3\overset{+}{H}N-CH(leuc)-CO-NH-$	0.103	Cowgill, 1963
Tyrosylglycine	$-CO-NH-CH_2-COO^-$	$-\overset{+}{N}H_3$	0.074	Cowgill, 1963
Glycyltyrosine	$-COO^-$	$_3\overset{+}{H}N-CH_2-CO-NH-$	0.070	Cowgill, 1963
Tyrosine	$-COOH$	$-\overset{+}{N}H_3$	0.056	Cowgill, 1963
Glycyltyrosylglycinamide	$-CO-NH-CH_2-CO-NH_2$	$_3H\overset{+}{N}-CH_2-CO-NH-$	0.035	Cowgill, 1963
Tyrosylalanine	$-COOH$	$_3H\overset{+}{N}-CH_2-CO-NH-$	0.09	Cowgill, 1963A
Tyrosylphenylalanine			0.08	
Tyrosyltyrosine			0.08	
Copolymer n(L-glut-L-tyr)				Pesce et al., 1964
pH 7, 0.2 M NaCl			0.038	
The same, pH 3.0 (α-helix)			0.08	
n(DL-glut-L-tyr) pH 7 (coil)			0.038	
The same, pH 3 (helix)			0.069	
n(L-lys-L-tyr) pH 7			0.09	
n(L-lys-L-tyr-L-lys) pH 7			0.020	

Table 6. Dependence of Quantum Yield of Fluorescence of Some Proteins on Their Content of Free COO^- Groups (Teale, 1961)

Protein	Quantum yield of tyrosine fluorescence	$\frac{COO^-}{\text{tyrosine residue}}$
Zein	6.0	1.0
Insulin	3.7	1.14
Ribonuclease	1.7	1.4
Human serum albumin	1.0	1.65

increase in the quantum yield of fluorescence (Vladimirov, 1960).

In fact, in 5,3-ketosteroid isomerase, the fluorescence of which is due to four normally titratable (not forming a hydrogen bond) tyrosine residues, the fluorescence quantum yield is high and amounts to 65% of that of an equivalent tyrosine solution (Wang Shu-fang et al., 1963).

Thus, the combined action of two factors—the influence of electronegative groupings, primarily peptide, on the one hand, and the specific interaction with the neighboring charged carboxyl group of aminodicarboxylic acids, leading to partial ionization of the hydroxyl group of tyrosine, on the other—causes a reduction in the quantum yield of the fluorescence of class A proteins. Whereas the nature of the primary, secondary, and tertiary structure of various native proteins and changes in the secondary and tertiary structure in denaturation processes have practically no effect on the position of the maximum of the fluorescence spectra of class B proteins, the same cannot be said of the quantum yield.

Konev (1957) observed a 14% increase in the quantum yield of zein due to thermal denaturation, whereas Burshtein (1964) observed a 35% decrease in the quantum yield of insulin.

Thorne and Kaplan (1963) reported an increase in the intensity of the tyrosine fluorescence of swine heart malate dehydrogenase on treatment with 6.5 M urea.

Cowgill (1964) observed a doubling of the quantum yield when RNase was denatured by 8 M urea. The nature of the action of urea, mediated by conformational changes in the macromolecule according to the scheme—change in conformation → change in microenvironment of tyrosine residues → change in quantum yield—is indicated by the absence of this effect of urea on the tripeptide glycyltyrosylglycinamide. The action of urea is also indicated by the agreement between the concentration thresholds for the initiation of the action of urea on the luminescence and for denaturation (4 M) and by the similar behavior of the fluorescence intensity and the absorption difference spectra ($\lambda = 287$ mμ). The denaturation process, estimated from luminescence intensity, conforms well to first-order kinetics.

According to Cowgill (1964), not only urea, but other methods of destroying the native structure of the macromolecule, such as treatment with certain detergents (dodecyl sulfate) or the breakage of disulfide bonds, which cement the framework of the macromolecule by means of mercaptoethanol, also lead to an increase in quantum yield. It is characteristic that in the last case the increase in luminescence intensity is not due to the removal of the quenching effect of the SH groups closest to tyrosine but is due to structural alterations, since breakage of the adjacent disulfide bond in the polypeptide oxytocin, which does not have a secondary and tertiary protein structure, does not lead to an increase in quantum yield. Gally and Edelman's plots (1962) of the fluorescence intensity of ribonuclease against temperature in the transition region did not show any decrease in fluorescence intensity with temperature but, on the contrary, an increase. The transition temperature at pH 5 was 57 to 58°C, which agreed with optical rotation data (Gally and Edelman, 1964).

A thorough investigation of bovine pancreatic ribonuclease showed that 50% of the tyrosine residues have abnormal properties: They are resistant to attack by iodine, they have a low dissociation constant, and their absorption difference spectra are unaltered by various treatments (Scheraga and Rupley, 1962). An analysis of the experimental data led the authors to the conclusion that this half of the tyrosine residues is buried in a hydrophobic cavity within the protein molecule. This conclusion, however, seems at first glance inconsistent with the low quantum yield of the fluorescence of this enzyme, since a nonpolar environment should lead to the reverse effect. Cowgill (1964A) successfully removed this contradiction by showing that, in nonpolar hexane solutions, phenol easily enters into a complex with N,N-dimethyl acetamide according to the scheme

$$C_6H_5{-}OH + O{=}C\begin{smallmatrix}\diagup CH_3\\ \diagdown N(Me)_2\end{smallmatrix} \rightleftarrows C_6H_5{-}OH \cdots O{=}C\begin{smallmatrix}\diagup CH_3\\ \diagdown N(Me)_2\end{smallmatrix}$$

which leads to fluorescence quenching.

The phosphorescence of class A proteins, like their fluorescence, is due to tyrosine. Vladimirov and Burshtein (1960) recorded a structureless fluorescence band with a maximum at 390 mμ for zein in aqueous solutions. The lifetime of the excited triplet states of tyrosine in proteins is 2.1 sec (Longworth, 1961).

According to Douzou et al. (1961), histone has a phosphorescence maximum at 395 mμ and τ = 1.74 sec.

Konev and Bobrovich (1964) showed that heavy atoms (lithium bromide and iodide) facilitate triplet-singlet transitions of tyrosine in proteins. This leads to the appearance of an intense, well-structured phosphorescence of tyrosine residues with fine-structure maxima at 354, 367, 376, and 397 mμ. The elements of the fine structure of the phosphorescence reflect the vibrational levels of the ground state.

CLASS B PROTEINS (Tryptophan-Containing Proteins)

One of the main features of the luminescence of proteins of this class is that, despite their containing three different fluorescent amino acids—phenylalanine, tyrosine, and tryptophan—only the tryptophan maximum appears in the fluorescence spectra of these proteins (Vladimirov and Burshtein, 1960; Teale, 1960). Yet the luminescence of all three components is found in the fluorescence spectra of equivalent amino acid mixtures of these proteins. Hence, the fluorescence spectrum of the protein macromolecule is not an additive spectrum of the aromatic amino acids contained in it but is usually formed by the fluorescence of only one of them, tryptophan. This is very clearly revealed by an investigation of the protein fluorescence spectra in relation to the wavelength of the exciting light. As long ago as 1957, I investigated this relationship by the following method (Konev, 1957, 1958, and 1959): The fluorescence was excited by monochromatic light through an SF-4 spectrophotometer, and the spectral composition of the fluores-

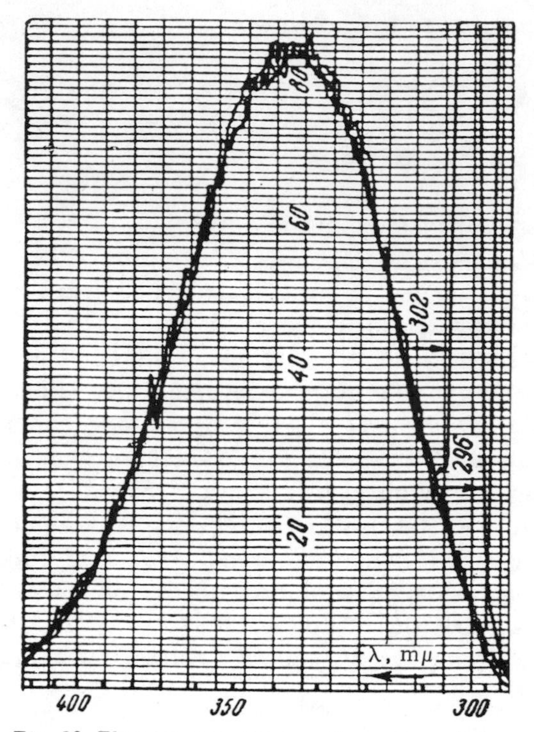

Fig. 28. Fluorescence spectra of aqueous solution of chymotrypsinogen excited by light of wavelength 253, 265, 296, and 302 mμ (Konev, Bobrovich, and Chernitskii, 1965).

cence was determined from a comparison of the relative intensities of the fluorescent light which passed through combinations of filters isolating different spectral regions.

The spectral composition of the protein fluorescence was practically unaltered when the wavelength of the exciting light was altered, i.e., when the relative excitation of phenylalanine, tyrosine, and tryptophan was altered. This conclusion was confirmed by recording the fluorescence spectra of proteins excited by monochromatic light on an apparatus consisting of two quartz monochromators and a photomultiplier cooled with liquid nitrogen (Fig. 28). Hence, the fluorescence spectra of most tryptophan-containing proteins are due to a single center—tryptophan—and tyrosine and phenylalanine are in a nonluminescent state in proteins (or, to be more precise, the contribution of tyrosine to the fluorescence spectrum of the protein is small in comparison with that of tryptophan).

This raises the question of why phenylalanine and tyrosine, which luminesce in the free state, are almost incapable of luminescence when they are incorporated in the polypeptide chain of a protein. Why is the fluorescence of these two amino acids quenched in proteins?

The most obvious explanation of the quenching was energy migration between phenylalanine and tyrosine, on the one hand, and tryptophan, on the other (Konev, 1957, 1958, and 1959; Steele and Szent-Györgyi, 1958; Vladimirov and Konev, 1959; and Vladimirov and Litvin, 1960). This idea was strengthened by the fulfilment of the donor-acceptor conditions for resonance migration of energy between these amino acids and by the fact that energy transfer to tryptophan had been recorded in mixed crystals of phenylalanine and tryptophan (Vladimirov, 1957) and of tyrosine and tryptophan (Vladimirov, 1961).

An investigation of fluorescence excitation spectra can help to reveal if tyrosine in fact acts as a sensitizer of the fluorescence of tryptophan in proteins. Vladimirov and Litvin (1960) and Teale (1960) carefully measured the fluorescence excitation spectra of several proteins at room temperature and low temperature (-160 to -170°C) in the range of 200 to 300 mμ. They found that the tyrosine component was almost absent in the excitation spectra and the fluorescence excitation spectra of the proteins was due to tryptophan. The screening, and not sensitizing, role of tyrosine and phenylalanine was confirmed by the increase in the quantum yield of the fluorescence at wavelengths 295 to 310 mμ, where these amino acids do not absorb (Teale, 1960). Hence, the excitation spectra and the absolute quantum yields of the fluorescence indicated that the nonfluorescent state of tyrosine and phenylalanine in proteins cannot be attributed to energy migration. It might still be postulated, however, that tryosine-tryptophan energy migration in proteins is such that its quantum yield is very low, i.e., that the energy of the electronic excited state of tyrosine is dissipated as heat in the act of energy migration.

Teale (1960 and 1961) subjected the tryptophan of human serum albumin to photooxidation by light of 254 mμ in the presence of methylene blue. Destruction of only the single tryptophan residue did not lead to the appearance of the fluorescence of the 17 tyrosine residues. In another experiment this protein was proteolytically fragmented by chymotrypsin and carboxypeptidase. This gave rise to protein fragments containing tyrosine, but no tryptophan. In this case again there was no emission of tyrosine fluorescence. Although a certain percentage of the quanta (2 to 10%) absorbed by tyrosine can migrate to tryptophan (this will be discussed more fully in Chapter 3), the main reason for the low quantum yield of the luminescence of tyrosine in proteins lies in the nature of the states of this amino acid and not in migration of energy.

As in the case of class A proteins, the reason for the quenching of tyrosine fluorescence in class B proteins is the ability of the hydroxyl group of the benzene ring to form a hydrogen bond with charged carboxyl groups or with the carbonyl group of the peptide bond. In the excited state the phenolic hydrogen is detached and tyrosine is ionized, which leads to the formation of the phenolic ion, which is incapable of luminescence:

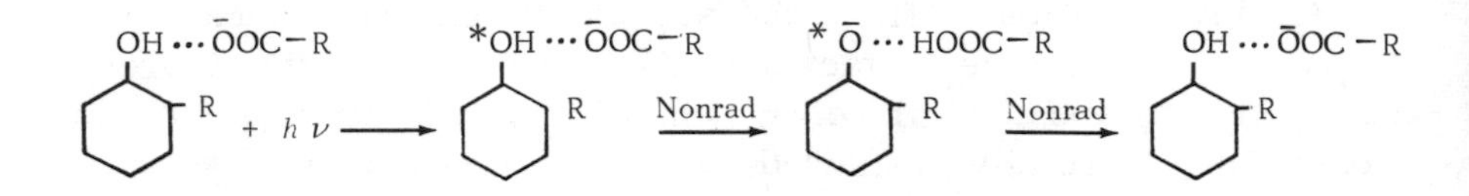

As mentioned above, tyrosine fluorescence is selectively quenched by compounds containing charged carboxyl groups (Teale, 1960). The formation of tyrosine-carboxyl hydrogen bonds increases the pK of ionization of the phenolic group (Laskowski et al., 1960). Measurement of protein titration curves shows that the pK of tyrosine ionization in them is actually shifted in the alkaline direction. For instance, in ribonuclease half of the tyrosine residues have pK 10.2 and half have pK 11.5; in ovalbumin all the phenolic groups of tyrosine have pK 11.5.

The bound state of tyrosine is indicated also by the absorption spectra. According to Crammer and Neuberger's data (1943), the appearance of the absorption maximum of ovalbumin in the 292-mμ region on ionization of the phenolic groups of tyrosine occurs at pH 12.5 instead of 10.

Two titration waves, normal and abnormal, were subsequently observed for many proteins, including chymotrypsinogen (Hermans, 1963), ribonuclease (Blumenfeld and Levy, 1958; and Hermans, 1963), myosin (Stracher, 1960), and lysozyme (Tanford and Wagner,

1954; and Donovan, Laskowski, and Scheraga, 1960). Tramer and Shugar (1959) came to the conclusion that tyrosine in proteins can exist in three different states: free tyrosine, which is in contact with the aqueous phase; tyrosine which is bound by hydrogen bonds and dissociates at pH 12 or after thermal denaturation; and tyrosine which is extremely tightly bound and plays an important role in maintaining the secondary structure of the macromolecule.

Finally, the nature of the nonfluorescent state of tyrosine in proteins was conclusively confirmed by the elegant experiments of Vladimirov (1961), who showed that the breakage of the hydrogen bonds of tyrosine with the carboxyl groups in the protein leads to an immediate increase in the quantum yield to almost the same values which tyrosine has in the free state.

Fluorescent Spectra of Proteins

The fluorescence spectra of different class B proteins are due to the same chemical center, tryptophan. Yet an investigation of the fluorescence spectra of various individual proteins confirms the view that each protein has a set of tryptophan residues which differ in their physicochemical state. This is manifested in the different positions of the maxima of the fluorescence spectra of different proteins (Table 7). The positions of the maxima of the fluorescence spectra vary in a fairly wide range, from 328 mμ (edestin) to 342 mμ (bovine serum albumin).

Table 7 shows a common feature of the fluorescence spectra of proteins in comparison with free tryptophan: their displacement through 10 to 20 mμ in the short-wave direction.

Several considerations indicate that the main reason for the difference in the positions of the fluorescence maxima of proteins is that they contain at least two forms of tryptophan residues in different proportions: tryptophan residues in a hydrophilic microenvironment on the surface of the globule and tryptophan residues in a hydrophobic environment in the interior of the protein. In fact, after treatment of proteins with 8 M urea, i.e., after all the tryptophan residues of the protein have been brought into an aqueous, hydrophilic environment, all the fluorescence spectra of proteins become the same and similar as well to the fluorescence spectrum of tryptophan in an aqueous solution (Teale, 1960, and Table 7). The reverse procedure can also be carried out, i.e., measurement of the fluorescence spectra of proteins after all the tryptophan

Table 7. Positions of Maxima of Fluorescence Spectra of Proteins, $m\mu$ (Teale, 1960)

Protein	Water or aqueous buffer solution	At liquid-nitrogen temperature	In 8 M urea	Propane 1:2 diol	Heat-denatured protein
Lysozyme	341		350	343	
Trypsin	332		350	335	
Trypsinogen	332		350	335	
Chymotrypsin	334		350	338	
pH 2.0	332*				339*
pH 4.7	334*				343*
pH 8.6	339*				345*
Chymotrypsinogen	331		350	336	
Human serum albumin	339, 338*	317*	350	340	330*
Bovine serum albumin	342		350	343	
Ovalbumin	332, 334*	320*, 326*	340	336	339*
Fumarase	335				
Carboxypeptidase in 10% acetate	340		348	340	
Pepsin	342		350	344	
Fibrinogen	337		350	340	
Edestin in 1 M NaCl	328		350		
Hemoglobin globin	335		348		
Urease, pH 5.0	330				
Actomyosin, pH 5.5	330*, 338*				342*, 350*
Human gamma globulin, pH 7.3	336*				341*
Casein	335†		345†		335†

*Burshtein (1964).
†Konev, Lyskova, and Saloshenko (1964).

residues have been brought into a nonpolar hydrophobic microenvironment (experiments performed by the author and I. D. Volotovskii). It is known (Imanishi et al., 1965) that in an acid solution of dodecyl sulfate proteins acquire a highly α-helicized conformation and are invested with a "skirt" composed of the hydrophobic parts of the detergent molecules. Under these conditions the differences in the fluorescence spectra of the investigated proteins disappear. All the proteins acquire the short-wave fluorescence maximum characteristic of chymotrypsinogen, 331 to 332 $m\mu$. Low values of the dielectric constant of the microenvironment of tryptophan residues in proteins are also produced by the freezing of aqueous solutions. Here the fluorescence maxima of the proteins take the same values, 320 to 323 mμ (see Table 10, p. 100).

Finally, Burshtein's experiments (1965 and 1966) showed that the spectrum of the protein fluorescence which is difficult to quench with potassium iodide (i.e., the fluorescence of the tryptophan residues in the center of the globule) has a maximum in the region of 330 mμ.

Thus, the main reason for the dependence of the tryptophan fluorescence spectra of proteins on their conformation is the change in the dielectric constant of the microenvironment of tryptophan residues. This agrees well with the conclusions made from an examination of the behavior of free tryptophan in the first chapter. In this case the dipole-dipole or ion-dipole interactions take place during the time in which the tryptophan molecule is in the fluorescent singlet state, since the absorption spectra of different proteins are the same.

However, the possible effects of the microenvironment on the tryptophan molecule in the excited state are not limited to those mentioned above. In certain conditions the tryptophan fluorescence spectrum of proteins undergoes radical changes of shape; it is converted from a single structureless band into a spectrum abounding in structural elements (Bobrovich and Konev, 1964). Working with swine pancreatic amylase crystals, these authors observed, in addition to the usual fluorescence spectrum of aqueous solutions of the enzyme with a maximum at 328 mμ, an abnormal fluorescence spectrum of the enzyme crystals at liquid-nitrogen temperature with a main maximum at 314 mμ and distinct maxima at 288, 293, 300.5, 307, 325, and 340 mμ (Fig. 29). Yet this unusual fluorescence for proteins was accompanied by the perfectly ordinary phosphorescence spectrum characteristic of tryptophan. It is quite obvious that, in view of the unusual shape of the low-temperature fluorescence spectrum, special attention must first be paid to showing that it belongs to tryptophan and not to other luminescence centers. The tryptophan nature of the low-temperature fluorescence spectrum of crystalline amylase emerges from a whole group of facts.

1. The excitation spectrum of this luminescence is similar to the excitation spectrum for tryptophan, the only difference being a slight shift through 3 to 4 mμ in the short-wave direction.

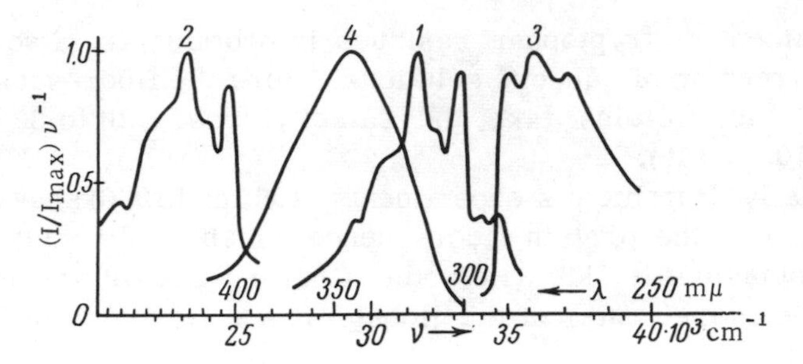

Fig. 29. Spectra of fluorescence (1), phosphorescence (2), and excitation of fluorescence (3) of amylase crystals, and luminescence of tryptophan crystals (4) at t = -196°C (Bobrovich and Konev, 1964).

2. The tryptophan indole ring, which is responsible for the spectral luminescent properties as previously mentioned, is capable in nonpolar solvents of producing a structured fluorescence band. The low-temperature fluorescence spectrum obtained for indole in cyclohexane is similar to the fluorescence spectrum of amylase.

3. The fluorescence spectrum of amylase is similar to its phosphorescence spectrum, except for the two short-wave maxima at 288 and 293 mμ. Hence, the system of energy levels of the ground state of the tryptophan molecule is manifested in the fluorescence of amylase.

4. Drying of amylase crystals extracted from the mother liquor or their solution in water leads to a gradual conversion of the abnormal fluorescence spectrum to the ordinary phosphorescence spectrum without any changes.

5. The shape of the fluorescence spectrum of amylase crystals does not depend on the exciting wavelength in the range of 255 to 296 mμ. This directly indicates that the fluorescence band is formed by only one luminescence center, not several. Apart from tryptophan, no other luminescence centers (such as tyrosine or phenylalanine) are implicated in the formation of this unusual fluorescence spectrum, since otherwise the shape of the fluorescence spectrum would depend on the wavelength of the exciting light.

Thus, all the structural elements in the amylase fluorescence spectrum shown in Fig. 29 are due to tryptophan alone.

This naturally gives rise to the question of the causes of this very pronounced modification of the shape of the fluorescence spectrum of tryptophan residues. First of all, the fact that the unusual spectrum is converted to the usual form when the crystalline structure is destroyed (by drying in air or solution in water) indicates that the physicochemical environment of tryptophan residues in the crystal lattice of protein is different from that of the same macromolecule in solution. At the same time, the change in the spectrum cannot be attributed to the interactions between tryptophan molecules (mainly the formation of hydrogen bonds between imino groups), which occur when the characteristic tryptophan crystalline packing is formed. This is indicated by the fact that tryptophan crystals have a low-temperature luminescence spectrum which differs distinctly from that observed in amylase. The low-temperature luminescence spectrum of tryptophan crystals is a single structureless band with a maximum at 340 mμ. Another special feature of the spectrum of tryptophan crystals is the complete absence of the ordinary violet phosphorescence, detectable only by phosphoroscope measurements, with a maximum at 500 mμ. This is not found in amylase, which exhibits the usual intense blue-violet phosphorescence. Hence, according to the fluorescence and phosphorescence spectra, the state of tryptophan in the crystal lattice of the protein differs from that of tryptophan in its own crystal lattice.

Thus, the changes in the spectral properties of tryptophan in the composition of amylase crystals are due to crysallization of the protein molecules. In general this means that proteins in the form of a crystal and in solution are qualitatively different. It is difficult at present to say whether this is due to changes in the degree of intrapolymeric hydration, to the close contact and interpenetration of the polypeptide chains of the macromolecules, or even to reversible changes in the secondary or tertiary structure of the protein at the moment of crystallization. In any case, it can be inferred that tryptophan luminescence conveys some information regarding the state of the protein as a macromolecule as a whole and that crystallization causes changes, structural or physicochemical, in this state.

The conclusion that there are some differences in the conformation of the protein in solution and in a crystal emerges also from the data for tritium-hydrogen exchange in solution and in an insulin crystal (Praissman and Rupley, 1964). It is still not clear what specific factors can cause the appearance of such pronounced structure in the fluorescence spectrum of tryptophan residues. Some analogy should probably be sought between the appearance of the distinct structure in the fluorescence of amylase crystals and the Shpol'skii effect (1962). The appearance of a quasi-line electronic spectrum (Shpol'skii effect) requires the incorporation of the activator in a crystal lattice corresponding in size and geometry to the activator molecules, which ensures its rigid fixation and identical position. It can be inferred that a similar situation exists in the case of tryptophan residues surrounded by the mosaic of polypeptide and side chains of amylase in the crystal state. In view of this, the fluorescence spectra shown in Fig. 29 must be regarded as the molecular spectra of tryptophan itself, undistorted by interactions. This conclusion is borne out by the fact that the fluorescence spectrum of amylase is similar to the fluorescence spectrum of indole at low temperature in nonpolar solvents and to the phosphorescence spectrum of indole and tryptophan in polar and nonpolar solvents. The presence of two short-wave maxima in the fluorescence spectrum of amylase crystals, which are absent in its phosphorescence spectrum, can be attributed, as mentioned above, to the fact that these two maxima belong to the ^1L_b electronic transition in fluorescence. This conclusion is confirmed by measurements of the emission polarization spectrum of the fluorescence of amylase crystals (Bobrovich and Konev, 1964A). These measurements showed that the two short-wave maxima have much higher values of polarization (10%) than the rest of the spectrum (5%) (Fig. 30).

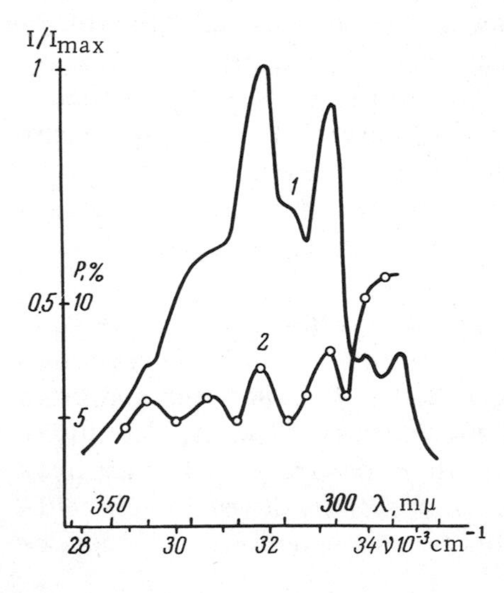

Fig. 30. Fluorescence spectrum (1) and emission polarization spectrum of the fluorescence (2) of amylase crystals, t = −196°C, with excitation at 265 mμ (Bobrovich and Konev, 1964).

Thus, the example of amylase in the crystalline state shows that the actual environment of tryptophan (the polypeptide chain and its different functional groups) can, by regular and induced dipole-dipole interactions, not only reduce the energy of the singlet excited state and smooth out the structure of the fluorescence spectra, but also lead to the opposite result. The rigid, congruent fixation of tryptophan residues in this case ensures the appearance of a tryptophan fluorescence spectrum which is similar to the molecular spectrum.

Quantum Yield of Fluorescence and Macrostructure of Proteins

The quantum yield of the fluorescence of tryptophan residues in proteins is sensitive to their macrostructural organization and usually has lower absolute values than aqueous solutions of free tryptophan ($Q \leq 0.2$). Values of the quantum yield of the fluorescence of various proteins are given in Table 8.

The very first hypothesis on the reasons for the variation of the fluorescence quantum yield of proteins was that concentration quenching may occur owing to intertryptophan energy migration. I put forward this hypothesis in 1957 to account for the changes in the fluorescence quantum yields of proteins due to denaturation and was supported by Vladimirov (1957), who discovered concentration quenching of fluorescence in aqueous solutions of tryptophan.

However, experiments in recent years have shown that the migration of energy between tryptophan molecules is not accompanied by concentration quenching. In a solid film of polyvinyl alcohol no concentration quenching was found up to a concentration of 1 M (Konev, Katibnikov, and Lyskova, 1963), i.e., up to concentrations which greatly exceed the actual concentration of tryptophan in proteins. The fluorescence of tryptophan at a concentration of 1 M was almost completely depolarized, which indicated that about 98% of the tryptophan residues are implicated in energy migration. In a similar way the dipeptide tryptophyltryptophan has a quantum yield of 0.09, i.e., the same as in dipeptides of tryptophan with acyclic amino acids (Cowgill, 1963A). Teale (1961) ascribed a special role to the charged amino group NH_3^+ in the quenching of tryptophan fluorescence. Cowgill (1963) attributed fluorescence quenching to the effect of the nearest peptide bond as an electronegative group.

As mentioned earlier, partial or complete removal of hydrogen

Table 8. Quantum Yields of Fluorescence of Tryptophan-Containing Proteins and Peptides (Teale, 1960; λ_{exc} = 280 mμ) (Q_1 by calculation for absorption of protein; Q_2 by calculation for absorption of tryptophan alone)

Protein (peptide)	Quantum yield				
	Water		Heat-denatured	8 M urea	
	Q_1	Q_2		Q_1	Q_2
Lysozyme	6.00	6.5		4.1	4.4
Trypsin	8.10	12.6		13.8	21.4
Trypsinogen	8.70	13.4		14.8	22.0
Chymotrypsin	9.50	10.5		20.4	22.6
Chymotrypsinogen	7.20	8.0		20.0	22.2
Human serum albumin	7.40	38.0	6.0 (pH 12)*	5.2	26.2
Bovine serum albumin	15.20	47.5		7.4	23.0
Ovalbumin	12.10	20.9	8.3*	13.1	22.4
Fumarase	9.00				
Carboxypeptidase	12.20	22.1		9.0	16.4
Pepsin	12.80	25.0		10.2	20.0
Fibrinogen	14.00	20.6		14.0	20.6
Edestin (1 M NaCl)	11.80	23.0		12.0	23.5
Globin	10.00	14.3		8.0	11.4
Glycyltryptophan	0.05†				
Alanyltryptophan	0.06†				
Leucyltryptophan	0.08†				
Prolyltryptophan	0.05†				
Phenylanalyltryptophan	0.06†				
Tryptophyltryptophan	0.09†				
Tryptophyltyrosine	0.12†				
Tryptophylglycine	0.14†				

*Vladimirov and Burshtein (1960).
†Cowgill (1963A).

from the imino group of the indole ring opens up an effective path for radiationless deactivation of the excited states of tryptophan. On the other hand, the fact that glycyltryptophan and acetyl tryptophan amide, which are regarded as "models" of the tryptophan residues in a protein, have smaller quantum yields than tryptophan and that proteins can have even higher quantum yields than tryptophan itself (casein granules in milk, Q = 0.5, Konev, 1965) suggest that interactions actually occur within the protein which lead to an increase in the fluorescence quantum yield of tryptophan residues.

It is tempting to link the fluorescence quantum yield with the existence in proteins of two forms of tryptophan residues in different proportions: tryptophan in a hydrophobic microenvironment inside the protein globule and tryptophan in a hydrophilic microenvironment on the surface of the globule. Judged by the behavior of glycyltryptophan and acetyl tryptophan amide, molecules in a weakly polar hydrophobic environment will have higher quantum yields than molecules in a polar aqueous environment. The existence of these forms of tryptophan is confirmed by data indicating quenching of protein fluorescence by lithium bromide (Konev, 1964) and potassium iodine (Burshtein, 1965) and further confirmed by the results of differential spectrophotometry when tryptophan residues in protein are gradually oxidized by hydrogen peroxide (Hoshijima et al., 1964). Steiner et al. (1964) managed to show that additives which reduce the dielectric constant of water (propylene glycol, sugar, dioxane, and dimethyl sulfoxide) enhance the fluorescence of acetyl tryptophan amide, and, conversely, additives which increase the dielectric constant (glycine and glycylglycine) quench the fluorescence. The initial slope of the curves relating the fluorescence intensity to propylene glycol concentration in the case of acetyl tryptophan was much greater than in proteins. This indicates that the tryptophan residues located inside the protein molecule are inaccessible for the solvent. However, it is sufficient to loosen the protein structure by solution of bovine serum albumin in 8 M urea or by tryptic digestion of thyroglobulin in 7 M urea, and the slopes of the curves for protein and acetyl tryptophan amide become similar.

Although the existence of these two forms of tryptophan in proteins is indisputable, we think that it is incorrect to assert that the hydrophobic form gives a higher fluorescence yield than the hydrophilic form. In fact, judged by the strong short-wave shift of the fluorescence band, trypsin, trypsinogen, chymotrypsin, and chymotrypsinogen contain the greatest relative amount of hydrophobic tryptophan. Yet the fluorescence quantum yield of these proteins is not high at all, about 10%, and increases to 22% only after treatment with 8 M urea, i.e., after the tryptophan residues are converted to the hydrophilic state. The high efficiency of radiationless quenching in these proteins is confirmed also by the low values of the τ of fluorescence (Table 9). This gives rise to a

Table 9. The Values of τ, Quantum Yields Q, and Positions of Maxima λ_{max} of Fluorescence of Some Proteins (Pikulik et al., 1966)

Protein	Q in water				Q in 8 M urea			
	$\tau \cdot 10^{-9}$ sec	Q, %	λmax, mμ	τ/Q·10	$\tau \cdot 10^{-9}$ sec	Q, %	λmax, mμ	τ/Q·10
Pepsin	4.5	25.0	342	1.80	3.6	20.0	350	1.80
Pepsin pH 9.0					4.0			
Chymotrypsin	3.0	10.5	334	2.85	3.7	22.6	350	1.64
Chymotrypsin after thermal denaturation	2.5				3.3			
Chymotrypsinogen	1.6	8.0	331	1.98	3.4	22.2	350	1.53
Hemoglobin globin	3.0	14.3	335	2.08	3.3	11.4	348	2.85
Hemoglobin globin after thermal denaturation	2.7							
Trypsin	2.4	12.0	332	2.00	3.5	21.4	350	1.63
Human serum albumin	4.1	22.0	339	1.90		26	350	
Casein	3.7							
Cytochrome C	3.5							
Edestin in 1 M NaCl	3.0	23.0	328	1.30	...	23.0	350	
Tryptophan	3.0	20.0	348	1.50	3.5			

contradiction: On the one hand, weakly polar solvents increase the fluorescence intensity of acetyl tryptophan amide and tryptophan residues themselves in proteins, and, on the other hand, tryptophan residues in hydrophobic regions have lower yields than in hydrophilic regions.

This became particularly clear after our experiments (Konev and Volotovskii), which showed that the fluorescence quantum yields of various proteins were reduced and leveled out at about 0.08 when all the tryptophan residues were brought into a hydrophobic microenvironment (1.0% solution of sodium dodecyl sulfate at pH 3). The contradiction can possibly be removed in the following way: Reduction of the dielectric constant of the medium, which increases the quantum yield of tryptophan derivatives with strongly quenched fluorescence (aqueous solution of glycyltryptophan), reduces it in compounds which have high initial yield values (indole and tryptophan). Yet, as I showed, the quenching effect of the peptide bond (glycyltryptophan) is greatly reduced in a viscous medium (glycerol or solid films of polyvinyl alcohol). Hence, in the case of tryptophan residues rigidly fixed in the protein structure and with fairly high quantum yields in a hydrophilic environment (0.20 to 0.25 in a solution of 8 M urea), their transfer to a hydrophobic microenvironment will not lead to an increase but to a reduction in the quantum yield, as actually happens. In the above-mentioned experiments of Steiner et al. the increase in the fluorescence intensity of "loosened" proteins under the action of propylene glycol was due mainly to additional increase in the viscosity of the microenvironment of the tryptophan residues which were previously in a hydrophobic medium.

In view of the above, with other conditions equal, we can accept a fluorescence quantum yield of about 0.20 to 0.25 for hydrophilic tryptophan residues and of about 0.08 for hydrophobic residues (as in chymotrypsinogen, the quantum yield of which is not altered by detergent treatment and which, judged by the position of the maximum, has the most hydrophobic tryptophan in its composition). Under actual conditions the fluorescence quantum yield of protein probably reflects some mean value of the yields of all the tryptophan residues in various physicochemical states and represents the resultant of several kinds of constant or induced dipole-dipole or ion-dipole effects of the surrounding groupings on tryptophan in the lowest singlet electronic excited state. All this leads to a wide range of variation of the fluorescence quantum yields of protein tryptophan: from 0.05 in wool keratin to 0.5 in milk casein granules. This suggests that there will be a fairly close connection between the fluorescence intensity of proteins and their secondary and tertiary structure, which virtually predetermines the specificity and intensity of the interaction of tryptophan residues with the microenvironment. Below we shall consider in more detail the bulk of the data revealing such a connection and shall outline the information which luminescence gives about the structural organization of the protein macromolecule.

Conformation and Luminescence of Proteins

We obtained the first data on the effect of the secondary and tertiary structure of the macromolecule on the ultraviolet fluorescence of protein in 1957.

In an investigation of the intensity of the fluorescence of native proteins with their secondary and tertiary structure destroyed by thermal denaturation, we observed an increase in fluorescence intensity in the case of rabbit blood serum gamma globulin on excitation by wavelengths 254 to 290 mμ. Later (Konev, 1958 and 1958A) an increase in fluorescence quantum yield was observed for arachin and ovalbumin but not for pepsin, gliadin, zein, or casein. On the other hand, in the case of casein, alkaline denaturation beginning at pH 9 to 10 led to a decrease in quantum yield. Alteration of the fluorescence quantum yield by thermal denaturation was confirmed in the carefully conducted experiments of Vladimirov and Burshtein (1960). The effect of conformational changes due to high temperatures and pH changes was investigated by Steiner and Edelhoch (1961).

The most interesting observation of these authors was the discovery of a good correlation between the region of change of luminescence intensity and the region of conformational changes.

On plots of fluorescence intensity against temperature (Fig. 31), which were almost linear for the region where no structural changes occurred, the slope changed at the structural transition point, i.e., the fluorescence intensity changed in time without further temperature increase. On passage through the region of this transition, hysteresis was revealed; i.e., when the temperature was reduced again, the whole curve was displaced.

For gamma globulin of rat blood serum at pH 11.2 in 0.1 M potassium chloride, the slope of the curve $I_{fl} = f(t)$ became less and

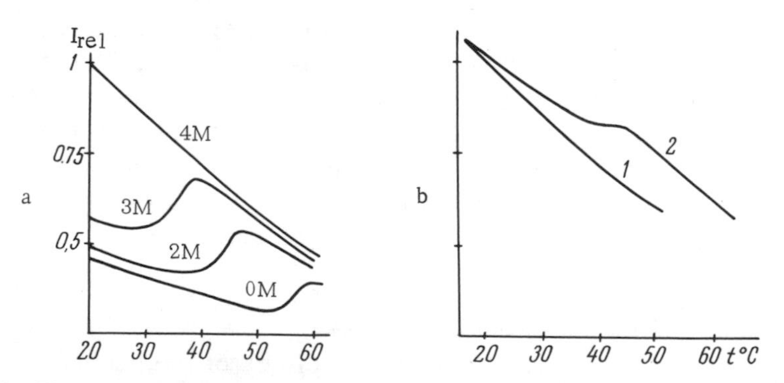

Fig. 31. Fluorescence intensity of proteins as a function of temperature: (a) Bence-Jones protein, phosphate buffer pH 7, and different concentrations of urea (Gally and Edelman, 1964); (b) chymotrypsinogen in water (1) and at pH = 1.9 (2) (Steiner and Edelhoch, 1963).

less negative above 35°C. On cooling, after the attainment of 60°C, the reverse branch of the curve lay above the straight line. For neutral solutions of the same antibody, which, according to biological tests, show no structural changes up to 60°C, the direct branch completely coincided with the reverse branch. The same result was obtained for lysozyme, for which there are no changes in the temperature range of 15 to 55°C and the pH range of 2.3 to 6.4 (Tanford and Wagner, 1954; Yang and Foster, 1955; and Donovan, Laskowski, and Scheraga, 1960).

Judged by tests which are independent of luminescence, chymotrypsinogen at pH 2 undergoes irreversible thermal denaturation at 35°C. In correspondence with this there is a distinct anomaly in the temperature quenching at pH 1.9: The structural change slows down the rate of quenching (Fig. 31a). For Bence-Jones proteins Gally and Edelman (1962 and 1964) discovered an anomalous temperature-quenching curve associated with a structural transition. In the region of the structural transition with a threshold temperature at 53°C there was no quenching but an increase of fluorescence intensity and a shift of the fluorescence maximum from 330 to 340 mμ. Distinct hysteresis was observed when the proteins were cooled again (Fig. 31b).

Heating of yeast alcohol dehydrogenase leads to a shift of the fluorescence maximum toward longer wavelengths and a reduction of the fluorescence intensity at temperatures of 36°C upwards. Changes in the luminescence parameters are correlated with a reduction of enzymatic activity (Brand, Everse, and Kaplan, 1962).

In the case of heart alcohol dehydrogenase the temperature threshold of the change in fluorescence intensity (60°C) coincides with the temperature thresholds for change in optical rotation and reduction of enzymatic activity.

Gerstein, Van Vunakis, and Levine (1963) compared one of the most sensitive tests for the tertiary structure of protein, the immunological antigen-antibody reaction, with a luminescent test. In their experiments antiserum to swine pepsinogen was produced in rats. When the pepsinogen was heated, its antigenic activity decreased in parallel with the reduction in fluorescence intensity.

Perlmann (1964) observed denaturation of pepsinogen which prevented its conversion to pepsin. The temperatures of the denaturation transition, determined from the increase in luminescence intensity, on the one hand, and from the optical rotation and conversion to pepsin, on the other, were similar to one another. The addition of urea in concentrations of 0.5, 1.0, 1.5, 2.0, 2.5, 3.0, and 3.5 M led to a gradual shift of the transition temperature toward lower values. This shift was the same in all tests. More thorough investigations showed that luminescence was more sensitive to biologically significant conformational changes than the optical rotation. After 15-min heating to 60°C followed by rapid cooling to 25°C, the optical activity was only 60% of its initial value, whereas the fluorescence and biological activity were completely restored. The luminescence intensity after cooling did not return to its initial value immediately, but gradually, with a rate constant of 23.1×10^{-2} min^{-1}. The same applied to the biological activity. Beginning at 50°C and pH 7, trypsin shows a structural transition to a very open state with a progressive increase in optical rotation and viscosity, reduction of relaxation time, increase in negative maxima at 291 and 286 mμ in the absorption difference spectra, and increase in fluorescence intensity. The structural transition conforms to simple first-order kinetics. The kinetic curves of the luminescence intensity coincide with those for the viscosity and the difference spectra (Steiner and Edelhoch, 1963).

The luminescence intensity reflects not only conformational changes due to heat but also structural changes in the macromolecule due to high urea concentrations and pH changes. Teale investigated the fluorescence quantum yields of many proteins in the native state and after denaturation with 6 M urea (Table 8).

For most proteins, trypsinogen, chymotrypsinogen, ovalbumin, fibrinogen, and edestin, the addition of 8 M urea leads to an increase in fluorescence quantum yield, but the increase is different for different proteins. For instance, the quantum yield of edestin increases by only 2%, whereas that of chymotrypsinogen increases by 180%. A small proportion of the investigated proteins show a reduction of quantum yield (lysozyme, carboxypeptidase, and pepsin), and the quantum yield of fibrinogen is unaltered by urea denaturation.

The properties of the tryptophan microenvironment, which are not the same in different individual proteins, become the same after urea denaturation, and the quantum yield takes values of 0.2 to 0.26, characteristic of tryptophan in the free state. The fluorescence maximum of most proteins is also stabilized at 348 to 350 mμ. The change in the quantum yield of the fluorescence due to denaturation can thus be used as an additional means of identification of individual proteins.

Velick (1961) observed very high values for the fluorescence quantum yield of lactate dehydrogenase (apparently about 0.7) and their reduction by a factor of 4 after treatment with 6 M urea. Gally and Edelman (1964) observed a structural transition in Bence-Jones proteins, accompanied by an increase in fluorescence intensity, on treatment with 8 M urea at less than room temperature.

In 9 M urea at pH values below 4.5 or more than 10.5 at room temperature, and at pH 7 at temperatures above 40°C, lysozyme undergoes progressive structural transitions. The luminescent kinetic curves of the transition coincide with the optical rotation curves and the difference spectra. The influence of the structure on the fluorescence intensity is very great. At pH 12.25 in 9 M urea and 0.01 M KCl at 25°C the fluorescence intensity increases by a factor of 2.5 in 8 to 10 min (Edelhoch and Steiner, 1962; and Steiner et al., 1964).

Pepsinogen undergoes a structural transition at room temperature in 4 to 9 M urea solutions at pH 6 to 7. This is accompanied by a progressive increase in optical rotation, by an increase in the negative peaks at 287 and 293 mμ in the absorption difference spectra, and by quenching (by 40%) of the fluorescence. The rate constant of this first-order reaction, calculated from luminescence measurements, agrees completely with the constants calculated from difference spectra and optical rotation data (Steiner et al., 1964).

Parallel changes in the fluorescence intensity and enzymatic activity are observed when lactate dehydrogenases of skeletal muscle, beef heart, and chicken heart are treated with increasing concentrations of urea in the range of 0 to 8 M (Brand, Everse, and Kaplan, 1962).

The rate constants of the structural transitions due to urea for lysozyme, pepsinogen, and soybean trypsin inhibitor are the same, according to measurements of the fluorescence, on one hand, and measurements of the optical rotation, viscosity, and difference spectra, on the other (Edelhoch and Steiner, 1964).

The effect of the third main method of denaturation, alteration of the pH of the medium on the fluorescence quantum yield, has also been investigated (Konev, 1957 and 1959; and Steiner and Edelhoch, 1963).

The results obtained for lysozyme are particularly illustrative. In aqueous solutions this enzyme does not undergo any change in conformation in the pH range of 2 to 12, and, hence, the curves of direct and back titration coincide, i.e., there is no hysteresis. In the pH range of 6.0 to 8.5 the quantum yield is constant, drops a little on the acid side, and is slightly higher on the alkaline side above pH 8.5 (Steiner and Edelhoch, 1963).

The situation is different if the enzyme is titrated in 9 M urea in the presence of 0.01 M potassium chloride. In an alkaline medium under these conditions there is a pronounced increase in fluorescence intensity, which becomes spontaneously greater with time. At pH 12.25 the intensity increases by a factor of 2.5 in 8 to 10 min, and at pH 2.75 the fluorescence intensity increases by 80%. It is most interesting that Edelhoch and Steiner (1962) managed to show complete parallelism in the change of the fluorescence intensity, optical rotation, and changes at the point 293 mμ in the absorption difference spectra, which reflect the pH dependence of these indices in direct and back titration. Hence, these experiments showed that changes in the fluorescence intensity directly reflect structural changes of the macromolecule.

A considerable fluorescence-quenching effect and hysteresis between the curves of direct and back titration were discovered in ovalbumin beginning at pH 12.6, in bovine serum gamma globulin beginning at pH 11.5, and in pepsin beginning at pH 7.5 to 8.5. As a rule, the structural transition points, identified from the luminescence as the threshold pH values at which hysteresis of the direct and back titration begins, agreed well with the transition points discovered by other methods: solubility, optical rotation, dye sorption, and enzymatic activity.

Thus, the two main luminescence tests—the temperature (or concentration) threshold for the initiation of changes in the luminescence intensity which are largely irreversible (hysteresis effect) and the kinetic curves of these changes—are quite consistent with other tests for structural changes of proteins—optical rotation tests, difference spectrophotometry, viscosity, and, what is most important, direct tests for biological activity. In several cases luminescence intensity is a more sensitive test for structural changes than other methods used in such cases. Three examples can be given. Soybean trypsin inhibitor (Edelhoch and Steiner, 1964) shows barely detectable changes in optical rotation, difference spectra, and polarization at acid pH, whereas the luminescence intensity changes more distinctly; it increases by 30%.

Bovine serum albumin in the pH range of 7 to 9 does not show any changes in optical rotation, although there are changes in the ability to bind calcium ions and the dye methyl orange. This suggested the existence of a structural transition under these conditions. In fact, Steiner and Edelhoch (1963A) found considerable quenching of the fluorescence of bovine-serum albumin at these pH values.

When pepsinogen is heated and cooled repeatedly, the optical activity is completely restored each time, although the biological activity gradually decreases. After five heating and cooling cycles, the biological activity is reduced to 70% of its initial value and, in complete correspondence with this, the fluorescence intensity is reduced to 64%. Hence, in this case the luminescence test is correlated better with the biological test than with the optical rotation test (Perlmann, 1964).

In other cases luminescence is less sensitive to conformational changes than are optical rotation, difference spectrophotometry, and viscosity.

Yet luminescence has one indisputable advantage over other methods of structural analysis, the possibility of working with extremely small amounts of protein. Moreover, luminescence measurements involve practically no time lag and are not greatly affected by the presence of other substances. This promises to provide a basis for the measurement of conformational changes in the living, functioning cell.

The foregoing indicates that protein luminescence is a very useful auxiliary method of investigating conformational changes in protein systems.

In recent years the idea of the dynamic nature of the secondary and tertiary structures of an enzyme during its functioning has been more and more widely accepted by enzymologists.

It has been suggested that the amazingly precise geometrical correspondence of the active enzyme center and the substrate does not exist beforehand but arises to some extent in the course of interaction of the enzyme and substrate [induced-fit hypothesis (Koshland et al., 1962)].

It has been suggested that (Eigen, Gordon, and Hammes, 1963) conformational changes in the macromolecule are required for a search for the best (critical) position of the substrate and coenzyme at which the bond to be broken is most effectively polarized and split. It is quite probable even that conformational changes themselves may cause the weakening and rupture of the bond (like a "loan" of energy at the moment of overcoming the activation barrier).

As the authors report, several types of conformational changes of serum albumin in a period of about 10^{-3} sec can be detected by relaxation methods. Eigen et al. (1964) managed to show that the helix-random coil transition time for polyglutamic acid and polylysine is less than 10^{-7} sec. This high rate of conformational changes makes their role in catalysis highly probable. It can be regarded as very remarkable that some of the earliest data indicating possible changes in the microenvironment of tryptophan residues in the protein macromolecule at the moment of "work" were obtained by the luminescence method in an investigation of the luminescence spectra of the urease-urea, enzyme-substrate complex, when Burshtein (1961) managed to show that the spectrum of this complex was slightly displaced in comparison with that of the "idle" enzyme. In other words, the macromolecular microenvironment of the tryptophan residues of urease in the act of enzyme catalysis is altered, i.e., there appears to be a change in the conformation of the whole macromolecule (or the active center). Soon after this, Sturtevant (1962) obtained results which indicated a change in the fluorescence quantum yield of α-chymotrypsin after the addition of several synthetic substrates and inhibitors to it.

A change in conformation of the enzyme macromolecule when the enzyme-substrate complex is formed has been shown by direct methods (optical rotatory dispersion) for chymotrypsin (Havsteen and Hess, 1962), for glutamic acid transaminase (Fasella and Hammes, 1964), and for D-amino acid oxidase in a complex with the coenzyme (flavin adenine dinucleotide) and in a triple complex with the coenzyme and an artificial substrate, benzoate (Yagi and Ozawa, 1962), etc.

Thus, the luminescence intensity reflects not only denaturing, largely artificial, conformational changes in the protein macromolecule, but also natural changes in the secondary and tertiary structure, which occur as part of the working mechanism of enzymes.

According to luminescence data, similar relationships are observed in the case of antigen-antibody interaction. The most interesting data in this connection have been obtained with botulinus toxin (Boroff and Fitzgerald, 1958; and Boroff, 1959). Botulinus toxin (*Clostridium botulinum* A, C, and D) is a protein with a high molecular weight (about 900,00), which readily breaks up into fragments with molecular weights of 45,000 to 70,000, i.e., in fact, into individual, independent protein molecules. Botulinus toxin contains 1.69% tryptophan. Transfer of the toxin to a 6 M urea solution leads to a reduction of approximately 40% in the luminescence intensity and to the complete loss of biological activity. The changes in quantum yield are probably due to breakup of the large botulinus toxin molecule into fragments (quaternary structure effect). In the complex of the two proteins (antitoxin-toxin), the quantum yield of the latter is much less than in the free state. This is not observed when the serum of a nonimmunized animal is combined with the toxin nor when antitoxin A is added to toxin C, or vice versa.

The conclusion of the authors regarding the identity of the active centers of toxicity and luminescence was subjected to a definitive examination by Schantz et al. (1960). These authors found that in several cases the toxicity of the preparation could be completely destroyed without appreciable change in luminescent power (in urea and guanidine) and that there was no correlation between the luminescence intensity and biological activity of chromatographic fractions of the crude toxin on cellulose. However, this criticism will be quite invalid if it is assumed that not all the tryptophan residues are in the active center, and, as for all enzymes, the toxin is biologically active only when the active center is present and the protein macromolecule as a whole is in its native state.

Finally, the third main way in which protein molecules function is by mechanochemical processes which transform the chemical energy of energy-rich bonds of ATP into mechanical work. The first luminescence investigations of the protein macromolecule during such functioning were carried out by Burshtein and Suslova (1964) and Shtrankfel'd (1964).

Using a continuous-flow cuvette Burshtein (1964) and Burshtein and Suslova (1964) investigated the changes in luminescence during the formation and breakdown of a myosin-ATP complex according to the scheme

$$E + S \rightleftarrows ES \rightarrow E + P$$

The formation of the complex was accompanied by a reduction in tryptophan fluorescence intensity, and the kinetics of the change in fluorescence intensity showed two distinct phases: a reduction of fluorescence intensity initially, reflecting the formation of the complex, and its gradual decay to the initial level, corresponding to the degradation. The difference spectrum for excitation of myosin fluorescence in the complex and in the free state $\Delta F_{max} = f(\lambda)$ had maxima at 274, 283.5, and 291 mμ, which were imposed on the steady decrease in fluorescence intensity of the complex. These maxima corresponded well with the maxima of the absorption difference spectrum of tryptophan at 274, 283.5, and 291 mμ. A characteristic feature was the change in shape of the kinetic curves of the protein luminescence with the exciting wavelength. This may be indicative of as yet undiscovered stages in the enzymatic reaction, which give a spectrally dependent contribution to the general picture.

Thus, the secondary and tertiary structure of the protein macromolecule has an appreciable effect on the fluorescence quantum yield of the tryptophan residues contained in it and on their excitation spectrum. This last fact means that the absorption difference spectra in the case of denaturating conformational changes of the protein can be replaced by the fluorescence excitation difference spectra, which are suitable for practical work with trace amounts of substances.

The luminescence intensity is apparently affected not only by the secondary and tertiary structures, but also by higher levels of structural organization. Konev and Lyskova (1963), Konev and Saloshenko (1963), and Konev (1964) showed that the formation of quaternary structures (granules of milk casein) leads to approximate doubling of the fluorescence intensity. The degradation of these granules by various chemical (urea, ammonium sulfate, and dodecyl sulfate) and physical (high dilution and ultrasound) factors leads to a reduction of luminescence intensity.

Lifetime of Excited State τ of Protein Fluorescence

The τ of fluorescence of several proteins is given in Table 9 (Pikulik, Kostko, Konev, and Chernitskii, 1966). The table shows that out of eight investigated proteins only three have the same values of τ — globin, edestin, and chymotrypsin. Hence, the τ of fluorescence can be regarded as an individual characteristic of the macromolecule. An investigation of the τ of fluorescence once again confirms the effect of the structure of the macromolecule on the state of tryptophan in the protein. Denaturation of proteins by 8 M urea leads to equalization of the τ of protein fluorescence, as is also the case for the quantum yields. The τ of protein fluorescence becomes equal to the τ of tryptophan fluorescence in urea, 3.5×10^{-9} sec.

On the other hand, when proteins are transferred to a 0.4% solution of dodecyl sulfate at pH 3, the τ values become equal at a "low" level: τ takes values of 1.6 to 1.9×10^{-9} sec.

What has been said in the last four sections of this chapter can be summed up as follows: Proteins contain at least two forms of tryptophan residues, which differ considerably in their scalar luminescent characteristics. Tryptophan residues in a hydrophilic microenvironment have high quantum yields (0.20 to 0.25) and large τ of fluorescence (3.3 to 3.7×10^{-9} sec), and the maximum of the fluorescence spectra lies in the long-wave region (350 mμ). The tryptophan residues in the hydrophobic microenvironment in the interior of the globule have, on the other hand, low quantum yields (0.08 to 0.11) and low τ of fluorescence (1.6 to 2.0×10^{-9} sec), and the maximum of the fluorescence spectrum is in the short-wave region (331 to 332 mμ). The secondary, tertiary, and quaternary structures of the protein macromolecule are linked with its luminescence parameters through changes in the proportion of hydrophilic and hydrophobic forms of tryptophan residues.

Absorption Polarization Spectra of Fluorescence of Proteins

The polarization of the fluorescence of proteins, measured by Weber (1960) and a little later, but independently, by Konev and Katibnikov (1961), was found to be much lower than in the case of free tryptophan (see Table 15).

Degrees of polarization much lower than for free tryptophan were observed not only in solutions of proteins in water but in a 50% binary mixture (propylene glycol-water) at -70°C (Weber, 1960) and also in proteins in the solid state at room temperature and at liquid-nitrogen temperature (Konev and Katibnikov, 1961), in proteins in glycerol and solid films of polyvinyl alcohol (Konev, Katibnikov, and Lyskova, 1964), and in casein contained in a solid milk granule and casein film (Konev and Lyskova, 1963; and Konev and Saloshenko, 1963). This indicates the migrational nature of the depolarization of protein fluorescence.

Absorption polarization spectra of the fluorescence of proteins have been investigated by Weber (1960, 1961, and 1963), Konev and Katibnikov (1961), Konev, Katibnikov, and Lyskova (1962 and 1964),

Konev, Bobrovich, and Chernitskii (1964), and Bobrovich and Konev (1964). In general, the shapes of polarization spectra of the fluorescence of proteins and tryptophan agree with one another, apart from the above-mentioned reduction in absolute values of the polarization of all proteins in comparison with free tryptophan (Fig. 32).

The minimum of the polarization spectrum in the region of 290 mμ indicates that the tryptophan residues contained in the protein absorb light with the participation of the same two electronic transitions $^1L_a \leftarrow A$ and $^1L_b \leftarrow A$ as in free tryptophan. In some proteins the tryptophan residues have a certain preferential orientation of their axes along one of the axes of the macromolecule. In such cases the protein may be oriented and the tryptophan residues will then also be oriented. This creates conditions for polarization detection of the absorption of the negative oscillator, as was described in the section devoted to tryptophan polarization spectra. In fact, it has been shown for wool keratin and silk fibroin that the negative ^1L_b oscillator forms an absorption spectrum which has a maximum at 290 mμ and is similar in general features to that of free tryptophan (Fig. 33) (Konev and Katibnikov). In several proteins (human and bovine serum albumin, pepsin, chymotrypsinogen, rat-muscle lactate dehydro-

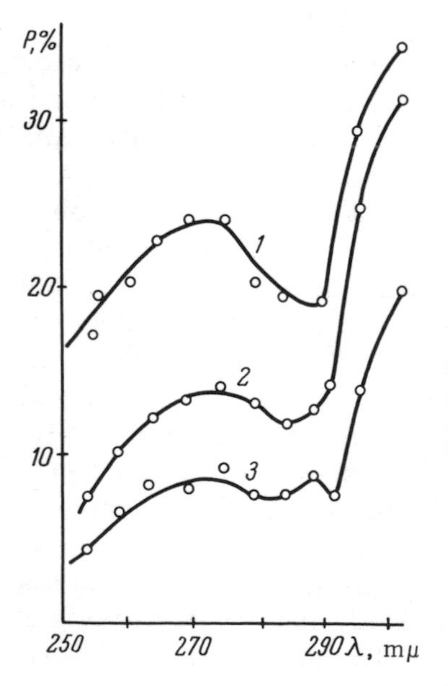

Fig. 32. Absorption polarization spectra of fluorescence of tryptophan in polyvinyl alcohol (1) and of human serum albumin (2) and chymotrypsinogen (3) in water, t = 20°C, recorded at 345 mμ (Bobrovich and Konev, 1965).

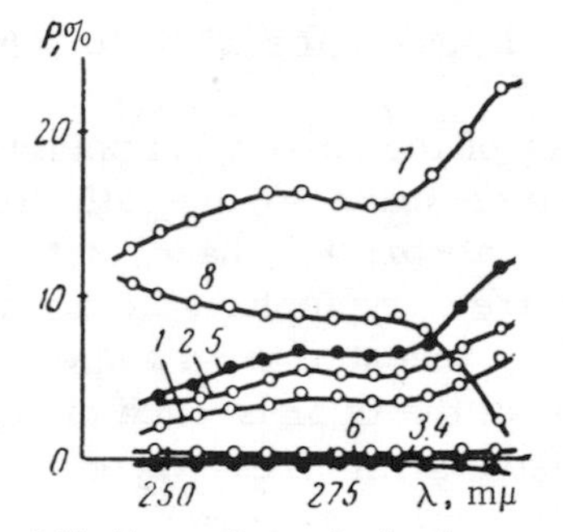

Fig. 33. (1 and 3) Absorption polarization spectra of fluorescence of proteins for randomly oriented wool keratin fibers, excited by light with the vibrations of the electric vector E in the vertical and horizontal directions, respectively; (2 and 4) the same as 1 and 3, but for randomly oriented silk fibroin fibers; (5 and 6) the same as 1 and 3, but for vertically oriented wool keratin fibers; (7 and 8) the same as 1 and 3, but for vertically oriented silk fibroin fibers. (Data obtained by the author in collaboration with M. A. Katibnikov.)

genase and lysozyme) the vibrational structure of the ^1L_b transition is revealed, as in the case of N-glycyltryptophan in propylene glycol at -70°C and in the case of indole in sugar crystals at room temperature. It consists of a second maximum at 285 mμ and a minimum at 292 mμ (Weber, 1960).

The height of this additional maximum, which reflects the microenvironment of the tryptophan residues, varies considerably from protein to protein and provides a fairly specific qualitative characteristic of various proteins. For instance, chymotrypsin does not have an additional maximum and minimum, but its precursor, chymotrypsinogen, has a polarization spectrum with distinct fine structure (Weber, 1960, 1961, and 1963).

Velick (1958 and 1961) observed high values of polarization of fluorescence (P = 0.22 to 0.25), independent of the exciting wavelength, for two enzymes, lactate dehydrogenase and glutamate dehydrogenase of liver. These values coincide with the polarization of free tryptophan in viscous media—0.25 in all cases. In view of the great difference between Velick's results and those of other authors, we are inclined to the view that there is some systematic error in his results. This is all the more likely in view of the fact that, according to Velick's data, the two enzymes have unnatural absorption polarization spectra of fluorescence in the form of straight lines in the 270- to 300-mμ region (Velick, 1958).

Thus, an analysis of the absorption polarization spectra of the fluorescence of proteins leads to the following main conclusions:

1. The long-wave absorption band of tryptophan residues in protein, like that of free tryptophan, is formed by two electronic transitions ^1L_a and ^1L_b, oriented at an angle to one another.

2. Depolarization of the fluorescence of tryptophan residues in protein in comparison with free tryptophan indicates energy migration.

3. The shape of the polarization spectra is sensitive to conformational features of organization of the macromolecules and to conformational changes due to, e.g., urea denaturation.

Emission Polarization Spectra of Fluorescence of Proteins

Emission polarization spectra of the fluorescence of proteins were obtained in 1964 by Konev, Bobrovich, and Chernitskii by using a photomultiplier cooled to liquid-nitrogen temperature and operating as a quantum counter (Vladimirov and Litvin, 1960). The polarization spectra of several proteins are shown in Figs. 38 and 18b. The emission polarization spectra of the fluorescence of proteins can be characterized by two main features: firstly, the constancy of the polarization (with a slight fall in the long-wave direction) throughout the main part of the fluorescence band, including its maximum; secondly, the increase in the polarization in the short-wave part of the fluorescence band at 310 to 295 mμ. This shape of spectrum indicates that the luminescence center is of a complex nature and that the short-wave part of the band is due to a different electronic transition from the rest of the band.

That the short-wave luminescence, which has a shorter decay time and hence is less depolarized, actually belongs to tryptophan and not to some other center is indicated by two facts. The fluorescence excitation spectrum at 300 mμ coincides with the fluorescence excitation spectrum at 340 mμ, on the one hand, and the whole fluorescence band of proteins is independent of the wavelength of the exciting light, on the other hand.

Hence, like tryptophan in the free state, proteins have two different singlet electronic excited states ^1L_a and ^1L_b, each of which in principle can participate in its own way in photochemical and hence photobiological process. In other words, the state of tryptophan residues in protein is more like the state of tryptophan in a weakly polar solvent of the polyvinyl alcohol type than that of tryptophan in a polar solvent of the glycerol type, for which, as was shown earlier, the whole fluorescence band is due to the single electronic transition $^1L_a \rightarrow A$. The fluorescence polarization spectra of individual proteins do not differ too greatly from one another. Only the absolute values of the polarization within the main band and the rate of increase of polarization toward the short-wave end differ. There is a definite qualitative relationship between these two characteristics of the polarization spectra. The lower the values of the polarization of the fluorescence in the main part of the band, the steeper the increase in polarization in the short-wave part of the polarization spectrum. The increase in the polarization is particularly sharp (from 3 to 12%) in the short-wave part of the spectrum in the case of wool keratin. The great reduction in polarization toward the long-wave end of the fluorescence band in such proteins as wool keratin (Fig. 18b, curve 3), where intertryptophan energy migration is most efficient, may be due to the fact that this part represents tryptophan residues which interact most strongly with the microenvironment. Hence, since they have the lowest fluorescence level, these residues are more likely to act as acceptors than donors in intertryptophan energy migration processes.

On the other hand, in human serum albumin (Fig. 34a, curve 5) there is no rise at all in the short-wave part of the polarization

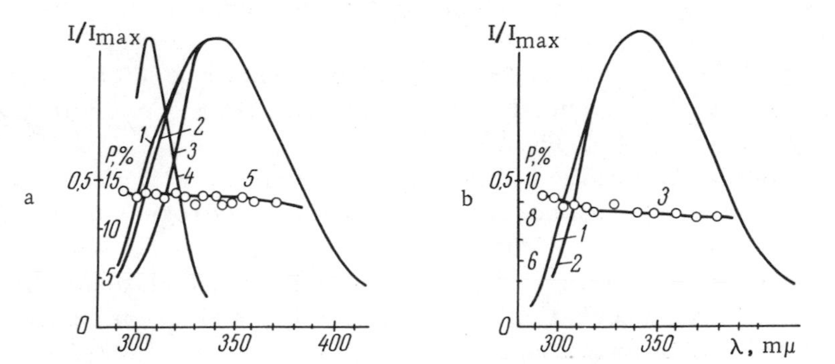

Fig. 34. (a) Aqueous solution of human serum albumin at t = 20°C: (1, 2, and 3) fluorescence spectra excited by light of wavelength 265, 253, and 296 mμ, respectively; (4) difference spectrum obtained by subtraction of fluorescence spectrum on excitation by 296 and 265 mμ; (5) emission polarization spectrum of fluorescence. (b) Aqueous solution of pepsin at t = 20°C: (1 and 2) fluorescence spectra obtained by excitation with 253 or 265 mμ and 296 mμ, respectively; (3) emission polarization spectrum (Bobrovich and Konev, 1965).

spectrum, and the polarization is constant at about 15% over the whole spectrum (Bobrovich and Konev, 1965). Monochromatic excitation of serum albumin by light with λ = 296 mμ gives a fluorescence spectrum with a distinctly "clipped" short-wave wing in comparison with the spectrum excited by 265 mμ, and the difference of these spectra corresponds to the fluorescence spectrum of tyrosine. Thus, in the extreme short-wave part of the protein spectra we are actually dealing with the fluorescence of tyrosine alone. Hence, the fluorescence of tyrosine, which is depolarized in this protein in comparison with the free state (15% instead of 35%), leads to straightening out of the polarization spectrum of the tryptophan fluorescence of proteins. In other words, in human serum albumin tyrosine-tyrosine energy migration, with efficiency of about 50%, occurs as well as tyrosine-tryptophan energy migration.

Phosphorescence of Proteins

It was initially believed that the phosphorescence spectra of proteins, like the fluorescence spectra, showed only the tryptophan component and that the tyrosine and phenylalanine components were quenched (Vladimirov and Litvin, 1960). Yet this rule, which holds in general, is not obeyed in several special cases. A distinct tyrosine maximum at 390 mμ is observed in the low-temperature luminescence spectra (77°K) of several proteins—aldolase, edestin, pepsin, and casein (Konev and Bobrovich, 1964)—on excitation by 265 mμ and disappears on excitation by 296 mμ, where tyrosine does not absorb.

Burshtein (1964) observed a tyrosine shoulder at 385 to 390 mμ when ovalbumin and human serum albumin were excited by light with wavelength 254 to 300 mμ.

Similarly, Augenstein and Nag-Chaudhuri (1964) observed a tyrosine shoulder in the case of alcohol dehydrogenase excited by 240 mμ; this shoulder was reduced when the exciting wavelength was changed to 280 mμ. According to their calculations, 0.5 of the total fluorescence on excitation by 280 mμ and 0.7 on excitation by 240 mμ is due to tyrosine.

Thus, for several proteins the contribution of the tyrosine component to the phosphorescence spectrum is considerable and is usually greater than the contribution of this amino acid to the fluorescence spectrum.

According to Longworth (1961), the phosphorescence at 388 mμ

for human serum albumin, lysozyme, and chymotrypsinogen has a decay time of 2.1 ± 0.1 sec and an excitation spectrum characteristic of tyrosine. The rest of the phosphorescence spectrum of these proteins has a decay time of 5.3 ± 0.1 sec and a tryptophan excitation spectrum.

The tyrosine component of the phosphorescence of proteins is increased by the action of factors causing the breakage of hydrogen bonds, which quench the fluorescence of tyrosine (Table 10). Thermal denaturation of albumin, casein, and edestin increases the luminescence at 390 to 400 mμ. A particularly intense, well-structured tyrosine phosphorescence is observed in proteins treated with 8 M lithium iodide. Fine-structure maxima of the phosphorescence of the tyrosine residues of several proteins (serum albumin, pepsin, etc.) are observed at 354, 366, 376, 387, and 397 mμ, i.e., at the same points of the spectrum as in the case of free tyrosine (Konev and Bobrovich, 1964). The intensity of these maxima is a fairly specific characteristic by which proteins can easily be distinguished from one another. For instance, the intense tyrosine maxima of human serum albumin are not found at all in aqueous solutions of chymotrypsinogen.

As in the case of fluorescence, there is no sign of phenylalanine in the phosphorescence and excitation spectra of proteins. Probably the only exception is a mixture of the tryptophan-free polypeptides, collagen and gelatin, for which Burshtein (1964) managed to discover a phosphorescence maximum at 420 mμ, which he ascribed to phenylalanine.

Thus, despite what has been said, tryptophan is still the main and, for some tryptophan-rich proteins, the only center responsible for the spectral phosphorescent properties of proteins in the natural state. The phosphorescence spectra, like the fluorescence spectra, indicate differences in the state of tryptophan in solution and in the polypeptide chain. This is manifested in a shift of the phosphorescence maxima through 5 to 7 mμ in the long-wave direction in comparison with free tryptophan in water.

When there is no interaction with the medium and the phosphorescence spectrum becomes similar to the molecular spectrum (amylase crystals at liquid-nitrogen temperature), the maxima of the phosphorescence spectrum of the protein practically coincide with those of tryptophan (405, 415, 430, 450, and 480 mμ). The much lower sensitivity of the phosphorescence to environmental influence in comparison with fluorescence becomes easier to under-

Table 10. Position of Low-Temperature Luminescence Maxima of Proteins and Relative Intensities of Fluorescence S and Phosphorescence T (Konev and Bobrovich, 1964)

Protein	Fluorescence, λ_{max}, mμ	Phosphorescence, λ_{max}, mμ				Relative luminescence intensity		
						S_{max}/T_{400}	S_{max}/T_{440}	S_{max}/T_{460}
Chymotrypsinogen								
native	322		412.5	439.0	450–465	4.8	1.80	2.3
denatured								
by heat	322		413.0	438.0	450–465	3.0	2.60	3.5
by 8 M urea	322		412.0	439.0	450–465	5.3	1.80	2.5
by 8 M urea + heat	322		412.0	439.0	450–465	3.5	1.80	2.4
Chymotrypsin								
native	323		412.0	439.0	450–465	7.0	2.20	3.3
denatured								
by heat	322		414.0	439.0	450–465	4.8	3.20	4.2
by 8 M urea	323		412.5	439.0	450–467	5.5	2.20	2.8
by 8 M urea + heat	323		410.0	436.0	450–465	4.5	2.00	2.6
Pepsin								
native	321	Shoulder 390	412.0	437.0	450–465	3.1	1.10	1.4
denatured								
by heat	320	Shoulder 390	412.0	437.0	450–465	3.1	1.10	1.5
by 8 M urea	320		409.0	435.0	445–465	3.2	1.20	1.5
Human serum albumin								
native	319	Shoulder 390	410.0	436.0	450–465	1.4	0.95	1.2
denatured								
by 8 M urea	317	Shoulder 390	407.0	435.0	450–465	0.8	0.76	1.2
by heat	322		410.0	435.0	450–465	1.4	1.00	1.3
Casein								
native	323	390	410.0	438.0	450–465	2.6	1.40	1.9
denatured by heat	323	390	414.0	440.0	450–465	1.8	1.20	1.6
Trypsin								
native	321		413.0	440.0	450–465	6.1	2.20	3.0
denatured								
by heat	322		412.0	437.5	450–465	3.0	2.20	3.3
by 8 M urea	322		411.0	437.5	450–465	4.1	2.00	2.4
by 8 M urea + heat	323		411.0	437.5	450–465	2.5	1.70	2.2

Table 10 (continued)

Protein	Fluorescence, λ_{max}, mμ	Phosphorescence, λ_{max}, mμ				Relative luminescence intensity		
						S_{max}/T_{400}	S_{max}/T_{440}	S_{max}/T_{460}
Ricin from castor-oil plant								
native	318	Shoulder 390	411.0	437.0	450–460	2.8	1.30	1.7
native, excited by 296 mμ	322		411.0	436.0	450–460	2.8	1.30	1.7
denatured								
by heat	318		408.0	435.0	450–465	1.2	0.85	1.5
by 8 M urea	318		409.0	435.0	450–465	2.3	1.20	1.8
Edestin								
native	318		411.0	437.0	450–465	3.5	1.30	1.7
native, excited by 296 mμ	322		411.0	437.0	450–465	3.5	1.30	1.7
denatured								
by heat	320		412.0	437.0	450–465	2.4	1.20	1.7
by 8 M urea	320		409.0	435.0	450–465	2.2	1.20	1.7
Aldolase								
native	320	Shoulder 390	413.0	438.0	450–465	2.1	1.20	1.8

stand when one considers that freezing makes the dielectric constants of different solvents more alike and rules out relaxation effects. Nevertheless, as Table 10 shows, even at 77° K there are some changes in the position of the phosphorescence maxima from protein to protein. This is seen most clearly for the 0–0 transition of the phosphorescence band (maximum at 410 mμ), which changes its position between 410 (human serum albumin and casein) and 413 mμ (aldolase and trypsin). Denaturation by heat or 8 M urea leads in several cases to some, though very slight, spectral changes in phosphorescence (Konev and Bobrovich, 1964).

Hence, the positions of the phosphorescence maxima of different proteins are very close to one another, which indicates that the shape of the phosphorescence spectrum is fairly insensitive to the influence of the secondary and tertiary structure of the macromolecule.

The ratio of the phosphorescence intensity to the fluorescence intensity behaves in a different way. Figure 35 shows that this ratio for amylase is six times higher than for chymotrypsin and that of the latter is 30% higher than for chymotrypsinogen, which has

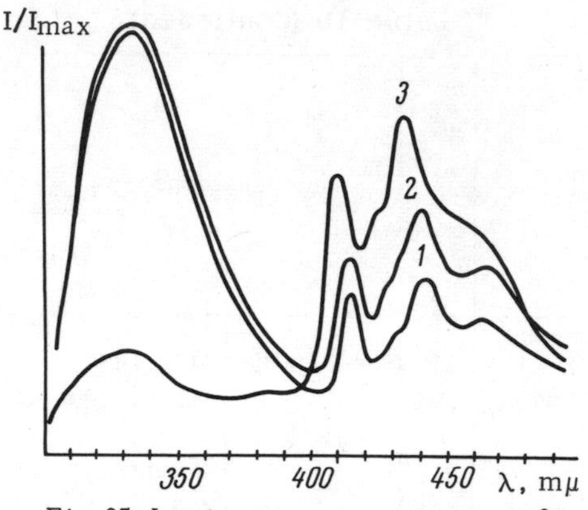

Fig. 35. Luminescence spectra at t = 77°K, pH 7.4, excited by 265 mμ: chymotrypsin (1), chymotrypsinogen (2), and amylase (3). (Date obtained by the author in collaboration with V. P. Bobrovich.)

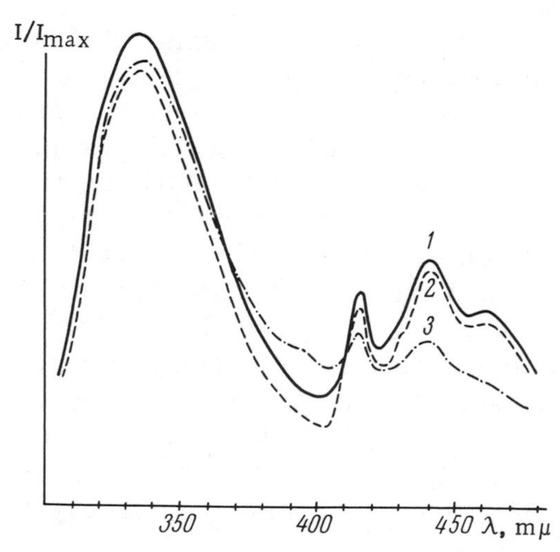

Fig. 36. Luminescence spectra at t = 77°K (phosphate buffer, pH 7.4, excited by 265 mμ) of chymotrypsinogen in the native state (1), after denaturation by 8 M urea (2), and after denaturation by heating to 100°C (3). (Data obtained by the author in collaboration with V. P. Bobrovich.)

a similar primary structure. Table 10 provides a fuller illustration of the range of variation of S/T.

The alteration of the secondary and tertiary structure of the protein macromolecule by thermal denaturation leads to changes in the relative phosphorescence intensity (Fig. 36). Proteins in which there is a slight superposition of tyrosine phosphorescence on the tryptophan phosphorescence show a strong relative reduction of the phosphorescence. In the case of chymotrypsinogen, it is 50% in comparison with the native state. In contrast to thermal denaturation, denaturation by 8 M urea has little effect on the S/T ratio for most proteins (Fig. 36; see also Table 12). In addition, prior urea denaturation of a protein protects it to some extent from the changes usually caused by thermal denaturation.

The reason that the probability of singlet-triplet radiative transitions is so sensitive to the macromolecular organization of the protein is most probably as follows:

As already mentioned, the distance of the proton from the nitrogen of the imino group has a strong effect on the relative phosphorescence intensity. In addition, since the secondary and tertiary structure of the protein creates a different microenvironment for tryptophan residues, the observed changes in luminescence intensity may reflect a different degree of involvement of the imino groupings in a hydrogen bond. In view of this we must assume for amylase, for instance, that the majority of tryptophan residues are completely free and are not bound with the neighboring substituents through the imino group. Changes in tyrosine phosphorescence intensity may also have some effect.

Decay Time of Phosphorescence of Proteins. The incorporation of tryptophan into the protein macromolecule has little effect on the decay time of the phosphorescence. According to Debye and Edwards (1952), the nature of the decay of protein phosphorescence is determined by two components. One, which decays exponentially, is the independent luminescence proper from the triplet level. The other is a slower, nonexponential component, due to recombination of the primary products of the photochemical reaction. According to Longworth (1961), the phosphorescence of the tryptophan component of human serum albumin, chymotrypsinogen, and lysozyme decays with $\tau = 5.3$ sec, i.e., the same as for free tryptophan.

The kinetic curves of phosphorescence decay over four orders of variation of intensity were recorded in the experiments of Konev,

Chernitskii, and Volotovskii. The measurements showed that for all the investigated proteins, except chymotrypsinogen and chymotrypsin, in contrast to indole and tryptophan, there were some deviations from an exponential law. This could be due to superposition of the shorter-lived tyrosine phosphorescence on the tryptophan phosphorescence. Consequently, the τ of phosphorescence determined at the earliest stage of the decay process was different in different proteins, whereas the τ of phosphorescence determined for the second and third order of reduction of the luminescence intensity was almost the same (Table 11). This fact and also the insignificant effect of 8 M urea and boiling on the tryptophan phosphorescence of proteins indicate that this luminescence parameter is very insensitive to the conformation of the macromolecule.

Absorption and Emission Polarization of Phosphorescence of Proteins. The vectorial characteristics of protein phosphorescence have been investigated in our laboratory (Konev and Bobrovich, 1965).

Table 11. Decay Time of Phosphorescence ($t_{1/2}$ and τ) of Some Proteins (T = 77°K, λ_{exc} = 280 mμ, λ_{rec} = 440 mμ) (Data Obtained by the Author in Collaboration with E. A. Chernitskii and I. D. Volotovskii)

Substance	$t_{1/2}$, sec, in water	τ_{ph}, sec		
		In water	In 8 M urea	In water after 3-min boiling
Chymotrypsinogen	3.23	5.9	5.9	6.0
Chymotrypsin	2.33	6.0	6.0	5.6
Trypsin	2.59	6.3	6.0	6.1
Casein	1.60	6.2	6.0	6.1
Edestin*	2.31	6.2	6.0	6.1
Bovine serum albumin		6.4	6.3	6.4
Human serum albumin		6.0		
Pepsin	1.65	5.8		
Urease		6.1		
Silk fibroin†		5.6		
Wool keratin†		5.5		
Tryptophan‡		7.4		

*In 1 N NaCl.
†Solid specimens.
‡Ethyl alcohol.

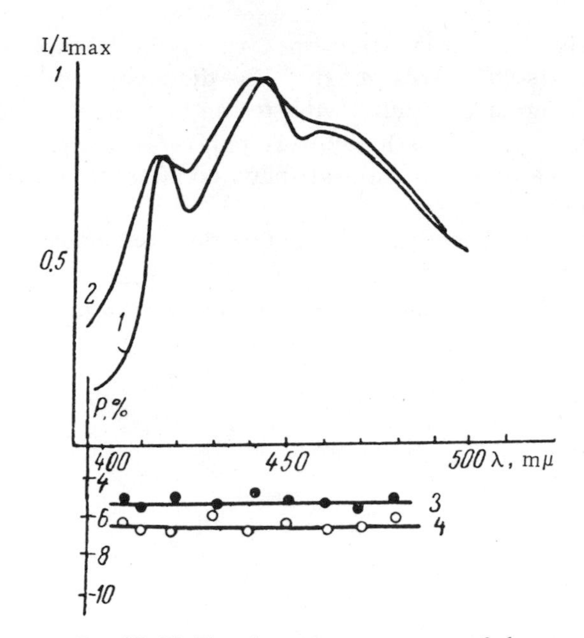

Fig. 37. (1) Phosphorescence spectrum of chymotrypsinogen in film of polyvinyl alcohol; (2) phosphorescence spectrum of nuclear fraction of white-rat liver in polyvinyl alcohol; (3) emission polarization spectrum of phosphorescence (chymotrypsinogen); (4) emission polarization spectrum of phosphorescence (nuclei), t = 196°C, by phosphoroscope with a time resolution of 10^{-3} sec (Konev and Bobrovich, 1965).

Protein phosphorescence has very low negative polarizations, about 4 to 7%. This depolarization relative to the free state of tryptophan phosphorescence again indicates intertryptophan energy migration in proteins. The polarization of phosphorescence is unaltered during decay, which indicates the impossibility of triplet-triplet energy migration. The portion of the decay curve in which the polarization was determined was isolated either by a phosphoroscope (with a time interval from 10^{-1} to 10^{-3} sec) or by an optical shutter, which allowed measurement of the polarization in a period of 0.2 to 10 sec after the cessation of excitation.

In view of the low polarizations, it is difficult to measure the absorption polarization spectrum of the phosphorescence accurately. Exploratory estimates indicate that it is similar to that of free tryptophan but has lower absolute values (film of aqueous solution of human serum albumin).

The emission polarization spectra of the phosphorescence of proteins consist of a straight line (the discovery of the vibrational structure is again difficult owing to the low absolute values) lying below the x axis (Fig. 37). Similar polarization spectra are given by proteins in their intravital physicochemical environment in nuclei and mitochondria.

All that has been said indicates that the oscillator model of triplet states obtained for free tryptophan can be extended to tryptophan residues in proteins and more complicated biological structures.

Chapter 3

MIGRATION AND CONSERVATION OF ENERGY IN THE PROTEIN MACROMOLECULE

The way in which the polypeptide chain is arranged in space, i.e., the secondary and tertiary structure of the protein, brings different chromophores very close together and leads to physical interactions between them. In the first place, the extremely dense packing of the material in the protein globule creates conditions for energy migration between energetically separate chromophores.

In speaking of energy migration processes in protein systems, we must distinguish between intramacromolecular energy migration and intermolecular energy migration, i.e., energy migration between the protein and some nonprotein molecule, a prosthetic group. Such prosthetic groups may be molecules of a vitamin, pigment, coenzyme, or dye. The range of questions associated with intermolecular energy migration has been discussed fairly fully in reviews by Vladimirov and myself (1957 and 1959). The most efficient energy migration process of the first type is intertryptophan energy migration.

It was found as far back as 1957 that there is no energy migration between sulfur-containing amino acids and tryptophan (Konev, 1957). My hypothesis (1957) of energy migration from peptide bonds to aromatic amino acids was not subsequently confirmed (Vladimirov and Litvin, 1960). Hence, delocalization of electronic excitation energy state in proteins mainly involves one amino acid, tryptophan.

The question of the existence of tryptophan-tryptophan energy migration in proteins has a rather peculiar history. In 1957, when we found an increase in the fluorescence intensity of several proteins after denaturation, we attributed the increase to reduction of migrational quenching of tryptophan fluorescence in the denatured protein due to some loosening of the macromolecule and an increase in the mean distance between the residues of this amino acid. These ideas appeared to be confirmed by Vladimirov's experiments (1957), in which concentration quenching of fluorescence in aqueous solutions of tryptophan was recorded. However, it became clear from subsequent work that changes in the fluorescence quantum yield of tryptophan could not be due to energy migration.

It is quite obvious that from scalar spectral characteristics it is difficult to detect the spatial separation of the light-absorbing center and the luminescence-emitting center for perfectly identical donor and acceptor molecules. The detection of energy migration in this case is assisted by the investigation of vectorial characteristics, fluorescence polarization. In this case energy migration is revealed by the different orientation of the tryptophan molecules which absorb and emit light, i.e., by fluorescence depolarization.

The polarization of the fluorescence of free tryptophan and proteins was measured by Weber (1960) and, independently of him, by Konev and Katibnikov (1961). By exciting the fluorescence with natural unpolarized light and working with thin (30-mμ) layers of tryptophan and proteins in propylene glycol at -70°C, Weber found that for 265 mμ the polarization of the fluorescence of tryptophan was 15%, whereas for proteins it did not exceed 6 to 8%. On conversion to linearly polarized exciting light the figures become 26 and 10 to 15%, respectively. Using linearly polarized light for excitation, Konev and Katibnikov (1961) obtained P_{265} = 25% for a glycerol solution of tryptophan and 2 to 7% for structural proteins of the keratin type of different origin (sheep's wool, calf's hair, and hen's quill).

Simple calculations and considerations ruled out the possibility of depolarization of the fluorescence due to secondary effects, i.e., light scattering and secondary luminescence. For instance, depolarization occurs in thin layers of protein where light scattering is practically absent.

More serious objections to an unambiguous conclusion regarding energy migration from the fact of fluorescence of depolarization stem from the possible alteration of the degree of symmetry of the emission oscillator of tryptophan incorporated in protein. If it is assumed that the tryptophan fluorescence oscillator is converted from the linear type to a completely or partially planar type, this in itself will lead to depolarization. This possibility was examined by Konev and Katibnikov (1961) and was rejected, since highly polarized fluorescence (up to 30 to 50%) could be obtained by orientation of tryptophan residues in proteins of the wool keratin or silk sericin type.

In 1960, however, Weber indicated another way, not involving energy migration, by which the fluorescence of proteins might be depolarized. Weber's theory was as follows:

As has been mentioned several times above, the long-wave absorption band of tryptophan at 280 mμ conceals two electronic transitions oriented at an angle of almost 90° to one another. The absorption spectrum of the negative oscillator at 289 mμ is inscribed within the broader absorption spectrum of the positive oscillator. These two absorption oscillators emit energy mainly via the same ^1L_a fluorescence oscillator, which coincides in direction with the positive ^1L_a absorption oscillator. Hence, the polarization of the fluorescence, which actually depends on the angle between the absorption oscillator and the emission oscillator, becomes independent of the percentage of quanta absorbed by the positive (positive values of fluorescence polarization) and negative (negative values of fluorescence polarization) oscillators. If, for instance, the absorption of the negative oscillator is increased for any reason, this will lead, without any involvement of energy migration, to a reduction in the positive values of polarization of the total fluorescence, i.e., to depolarization. On the basis of these considerations Weber suggested that the incorporation of tryptophan in the protein polypeptide chain is accompanied by changes in the tryptophan oscillator system whereby a greater number of absorbed quanta are allocated to the negative oscillator.

This initial hypothesis of Weber's was verified in 1961 by the following fact: It was found that the fluorescence of human serum albumin, which contains only one tryptophan residue, is also depolarized in comparison with free tryptophan. In this protein, of course, intertryptophan migration is absent, and one must then assume that the incorporation of tryptophan residues into the protein structure leads to a relative enhancement of the negative oscillator and, consequently, to a reduction in the polarization. This fact dealt a serious blow to the possibility of a migrational interpretation for other proteins.

The second forceful argument which enabled Weber to reject intertryptophan energy migration in proteins was provided by White's results (1960) for acid titration of the fluorescence of bovine serum albumin. The fluorescence of this protein, which contains two tryptophan residues, has two inflections in acid titration, at pH 0.5 and 3.5. The first of these inflections is associated with the tryptophan residue in which the carboxyl group is blocked by incorporation in the peptide bond. The second tryptophan residue is located in the proximity of the carboxyl group of aspartic or glutamic acid. The conversion of the carboxyl groups of dicarboxylic amino acids from the COO^- to the COOH state by acidification leads to quenching of tryptophan fluorescence. Weber argues as follows. If there is effective energy migration between tryptophan residues, then quenching of the fluorescence of one of them would lead to quenching of the fluorescence of the second, since the energy would migrate from it to the first residue and there be dissipated as heat. Hence, in the case of energy migration there would be only one step in acid titration. Actually, however, there are two, which indicates against intertryptophan energy migration.

This was Weber's argument in brief.

We shall try now to analyze thoroughly all the evidence for and against energy migration in proteins and consider the mechanism of depolarization of protein fluorescence.

One of the main differences between the state of tryptophan in aqueous solution and in the protein molecule lies in the more hydrophobic nature of the microenvironment (lower polarity) in the latter case. This naturally gives rise to the suspicion: Will this reduction in polarity of the medium not be accompanied by an increase in the relative contribution of the ^1L_b transition to absorption? This question can be answered by a comparison of the absorption polarization spectra of the fluorescence of tryptophan in different viscous solvents. Since the negative oscillator has a distinct absorption maximum at 289 mμ, the increase in its contribution to absorption would be manifested in a deepening of the dip of the polarization spectrum at this wavelength. Yet the experimental data contradict this suggestion. The dips of the polarization spectra of tryptophan and indole in glycerol (a highly polar medium), sugar crystal (of medium polarity), and polyvinyl alcohol film (a weakly polar medium) are practically the same (Chernitskii and Konev) (Table 12).

Moreover, the fluorescence polarization spectra of proteins

Table 12. Ratio $K = P_{270}/P_{289}$ for Tryptophan and Indole in Various Solvents

Medium	$K = \frac{P_{270}}{P_{289}}$	
	Tryptophan	Indole
Glycerol (room temperature)	1.3	
Propylene glycol (t = -70°C)	1.5	2.1
Sugar crystal (room temperature)	1.3	2.5
Polyvinyl alcohol film (room temperature)	1.3	2.1

themselves have even a slightly smaller dip at 289 mμ than does free tryptophan. Another fact which fits in well with this conclusion is that in protein absorption spectra the ratios of extinctions at 280 and 290 mμ are not altered in favor of long-wave absorption in comparison with free tryptophan. In other words, the ^1L_b electronic transition of tryptophan in protein is not appreciably increased in comparison with tryptophan in the free state. This obviously contradicts Weber's hypothesis regarding the relative increase in absorption of the negative oscillator.

The fact that incorporation of tryptophan in a peptide bond does not lead to appreciable changes in the oscillator nature of tryptophan is also indicated by the equal values of the polarization of the fluorescence of tryptophan and the dipeptide glycyltryptophan (Fig. 18a, curves 3 and 4). More distant groupings than the nearest peptide bond apparently have a slight effect on the polarization spectra. This emerges from experiments in which destruction of the secondary and tertiary structure of the protein macromolecule by heat and dioxane did not lead to changes in the shape of the polarization spectra (Konev, Bobrovich, and Chernitskii, 1965).

Thus, there are a whole series of arguments which show that changes in the oscillator model of tryptophan residues in protein cannot be attributed to depolarization of the fluorescence of protein in comparison with that of free tryptophan. On the other hand, we can cite evidence to show that depolarization of protein fluorescence is of a migrational nature. The most direct evidence of intertryptophan energy migration in proteins is provided by the emission polarization spectra of protein fluorescence (Konev, Bobrovich, and

Chernitskii, 1965). This is due first of all to the fact, discovered by the authors, that intertryptophan energy migration in model conditions causes very characteristic changes in the shape of the emission polarization spectra of fluorescence. With increase in tryptophan concentration in polyvinyl alcohol films and, hence, increase in intertryptophan energy migration the concentration depolarization is different at different points in the emission spectrum: The polarization in the long-wave part of the spectrum falls fairly sharply, whereas the extreme short-wave end of the spectrum is depolarized very slightly (Fig. 18a). At high tryptophan concentrations the behavior of the polarization spectrum is the reverse of that at low concentrations: Instead of the usual reduction in polarization in the short-wave region there is an increase. What causes this change in the polarization spectrum with intensification of energy migration processes?

The following processes underlie the observed changes in the shape of the polarization spectra. As mentioned earlier, the negative oscillator is responsible for only an insignificant fraction of the luminescence of tryptophan in polyvinyl alcohol. Transfer of the bulk of the absorbed energy to the positive oscillator is equivalent to quenching of its own fluorescence. A direct consequence of quenching will be the reduction of the lifetime of the ^1L_b excited state.

Since the effectiveness of energy migration depends on the lifetime of the excited state, depolarization will decrease with reduction of this time, i.e., in the region of intrinsic luminescence of the ^1L_b oscillator the depolarization will be very slight.

As Fig. 18a shows, this is actually observed.

Thus, the shape of the emission polarization spectrum of fluorescence can indicate the occurrence of energy migration processes. It only remains now to compare the fluorescence polarization spectra of proteins with the corresponding spectra obtained for concentrated films of polyvinyl alcohol. A comparison of the emission polarization spectra of the fluorescence of proteins (Fig. 18b) shows that they coincide with the corresponding spectra for tryptophan under conditions where efficient energy migration occurs. This would not have happened if the depolarization of the fluorescence of proteins was due to a different cause than that of the depolarization of the fluorescence of concentrated tryptophan solutions, i.e., not to energy migration, but to increased absorption by ^1L_b. In the latter case the reduction in polarization in the long-wave part of the spectrum would be compensated by its increase in the short-wave part.

In view of the importance of this conclusion we shall dwell in rather more detail on the emission polarization spectra of protein fluorescence.

Since simulation of the state of tryptophan residues in protein by means of tryptophan molecules in different conditions makes sense only if the whole fluorescence band of the protein is actually due to tryptophan alone, it was essential to compare the fluorescence spectra of a protein solution excited by monochromatic light of different wavelengths. The exact similarity of these spectra for excitation by monochromatic light of any wavelength in the range of 250 to 300 mμ definitely indicates that the only luminescence center of these proteins really is tryptophan (Fig. 28). In fact, the presence of other luminescence centers besides tryptophan would inevitably lead to a change in the shape of the fluorescence spectrum at excitation wavelengths which correspond to the absorption maxima of this other center (tyrosine and phenylalanine).

The second possible nonmigrational reason for the change in the emission polarization spectrum of protein fluorescence is the superposition of secondary, tertiary, and higher orders of fluorescence on the primary fluorescence. If a mixture of different orders of fluorescence is actually observed, then in the extreme short-wave part of the fluorescence spectrum the contribution of the secondary fluorescence would be greater than in the whole long-wave part of the spectrum. Reabsorption of fluorescence, by withdrawing quanta only from the short-wave reabsorbable part of the spectrum, would thus lead to rapid reduction of the ratio of the intensities of the primary and secondary fluorescence with approach to the short-wave end of the fluorescence band and, hence, to complete depolarization.

In fact, however, the emission polarization spectra of the fluorescence of proteins show a directly opposite relationship. The polarization increases toward the short-wave end of the fluorescence band. In addition, a very elementary calculation shows that fluorescence of the second and higher orders can be practically neglected even for concentrations where the light is totally absorbed at maximum 280 mμ. A comparison of the absorption and fluorescence spectra shows that in these conditions not more than 5% of the fluorescence quanta are absorbed. The fluorescence intensity will be given by the following equation: $I = (1-\beta) + \beta\varphi(1-\beta) + \beta^2\varphi^2(1-\beta)^t$ etc., where β is the percentage absorption of primary fluorescence in the investigated solution; φ is the fluorescence quantum yield; $1-\beta$ is the intensity of the primary fluorescence reaching the detector; $\beta\varphi$ is the intensity of the secondary fluorescence; $\beta\varphi(1-\beta)$ is the intensity of the secondary fluorescence reaching the detector; $\beta^2\varphi^2(1-\beta)$ is the intensity of the tertiary fluorescence reaching the detector, etc.

Putting $\varphi = 0.2$ and $\beta = 0.05$, we find that the intensity of the secondary fluorescence is 1% of that of the primary fluorescence.

In addition, the characteristic shape of the fluorescence polarization spectra remains unaltered even for thin and dilute protein layers where there is no reabsorption (the optical density even at the absorption maximum at 280 mμ is 0.05, i.e., about 5%). The constancy of the shape of the emission polarization spectra of fluorescence in a wide range from $D_{280} = \infty$ to $D_{280} = 0.01$ is the best evidence of the true nature of the recorded spectral relationships.

Thus, the possibility of simulating the fluorescence polarization spectra of tryptophan residues in protein by the corresponding spectra of free tryptophan under conditions where efficient energy migration occurs is a strong argument in support of the occurrence of energy migration in proteins. Additional evidence of the occurrence of intertryptophan energy migration can be obtained from the polarization of the phosphorescence of proteins. The use of polarization of the phosphorescence of proteins as an index of energy migration is of great interest from two points of view. Firstly, the phosphorescence is shifted more than 100 mμ in the long-wave direction in comparison with fluorescence, and, hence, in this case secondary light-scattering effects are greatly reduced. Secondly, and mainly, the tryptophan phosphorescence oscillator, as was shown in the preceding chapter, is almost perpendicular to the plane of the indole ring, i.e., the phosphorescence oscillator is perpendicular both to the ^1L_a oscillator and to the ^1L_b oscillator. Hence, changes in the relative contributions of the oscillators to absorption will not effect the polarization of the phosphorescence. On the other hand, energy migration processes will depolarize the fluorescence and phosphorescence of proteins to the same extent as for tryptophan.

Measurement of the absorption and emission polarization spectra of protein phosphorescence shows that the phosphorescence is depolarized to the same extent as the fluorescence (Fig. 37). For instance, the polarization of the phosphorescence of proteins is about -5 to -7%, i.e., approximately half as much as that of tryptophan itself (-13 to -15%). The fluorescence of proteins is depolarized to the same extent as that of free tryptophan. Hence, an experiment which excludes the possible depolarizing effect of the negative absorption oscillator still reveals considerable depolarization, which can be produced only by energy migration.

Finally, the third main experimental method of choosing between the migrational and oscillatory mechanism of depolarization consists in showing that the relative role of the oscillators in absorption is revealed practically instantaneously, at the instant of light absorption, whereas energy migration, which occurs throughout the lifetime of the excited state, will be accompanied by progressive depolarization of the fluorescence. The fluorescence polarization for molecules which have spent the least time in the excited state will be higher than for molecules with the greatest lifetime: The polarization will decrease as the fluorescence decays. Consequently, gradual quenching of the fluorescence, which first removes the fluorescence of the long-lived molecules and then that of the more and more short-lived, will lead to an increase in the polarization if the initial depolarization is due to energy migration. If depolarization is due to a change in the relative absorption of the oscillators, such a relationship will obviously not be observed, and the polarization of the fluorescence will remain constant as the fluorescence decays.

As is known, S. I. Vavilov regarded depolarization of luminescence during decay as one of the most direct indications of the existence of energy migration.

Figure 38 shows that twentyfold quenching of the fluorescence of casein by fluorescein (solid films of casein with fluorescein) leads to an increase of 10 to 15 to 16% in the polarization of the fluorescence. Two very important conditions are fulfilled in casein films with strongly quenched fluorescence: The tryptophan part of the fluorescence spectrum is not distorted, and this part of the fluorescence spectrum is independent of the wavelength of the exciting light. The increase in the fluorescence polarization is uniform over the whole long-wave part of the fluorescence spectrum, as is indicated by the illustrated emission polarization spectra (Fig. 38). This experiment can be regarded as a crucial one, and conclusively indicates the occurrence of energy migration in

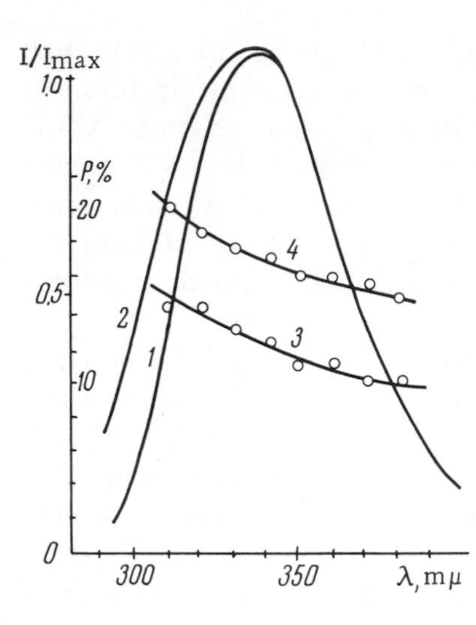

Fig. 38. Casein films, t = 20°C, excited by 265 mμ: fluorescence spectra without quencher (1) and quenched twentyfold by fluorescein (2), emission polarization spectra of fluorescence of unquenched (3) and twentyfold-quenched (4) casein (Konev, Chernitskii, and Muraveinik, 1965).

proteins, since no other factors (secondary effects, changes in oscillator model, etc.) can explain the increase in polarization with reduction of the lifetime of the luminescent centers. The possibility of rotational depolarization of long-lived molecules in this case is completely ruled out; the relaxation time of proteins, even in aqueous solutions, is 20 to 50 times greater than that of the excited state, to say nothing of peptides, where these times differ by a factor of hundreds of thousands. Direct experimental confirmation of this is the agreement of the fluorescence polarization values for solutions of proteins in water and glycerol at room temperature and of solid films of proteins (infinite viscosity) at room temperature and at liquid-nitrogen temperature.

One of the best demonstrations of the occurrence of intertryptophan energy migration in proteins was obtained in the case of hemoglobin (Konev, Bobrovich, and Chernitskii, 1965). As is known, efficient energy transfer from tryptophan to heme in hemoglobin leads to quenching of tryptophan fluorescence. According to our data, the quantum yield is 0.5%, and, according to Teale (1959), it is less than 0.2%. Hence, the excitation energy which is not transferred to heme will appear as luminescence with an extremely short decay time

$$\tau_1 = \frac{\tau_0 Q_1}{Q_0} = \frac{3.0 \cdot 10^{-9}\ \text{sec} \cdot 0.005}{0.14} \approx 10^{-10}\ \text{sec},$$

where τ_1, and τ_0 are the decay times, and Q_1 and Q_2 are the quantum yields of the fluorescence of hemoglobin and globin, respectively. In other words, the excited states of tryptophan which have been able to avoid participation in the highly efficient tryptophan-heme migration will be all the more likely to escape the considerably slower process of intertryptophan migration. In fact, measurements of the fluorescence polarization spectra of hemoglobin and its protein part, globin, completely confirm this hypothesis. The globin of hemoglobin exhibits typical protein absorption and emission polarization spectra with polarization values P_{265}^{350} of about 9 to 10% (Fig. 39). The attachment of the heme groupings does not alter the long-wave profile of the luminescence and the position of the maximum but increases the relative intensity of the short-wave part of the spectrum. This is another confirmation of the fact that the fluorescence is formed by the long-lived $^1L_a \rightarrow A$ luminescence, which undergoes efficient quenching, and by the short-lived $^1L_b \rightarrow A$ luminescence, which is less quenched by the attached heme (Fig. 39). At the same time, the luminescence of the ^1L_a oscillator left after quenching gives high polarization values. The absorption and emission polarization spectra of the fluorescence of hemoglobin

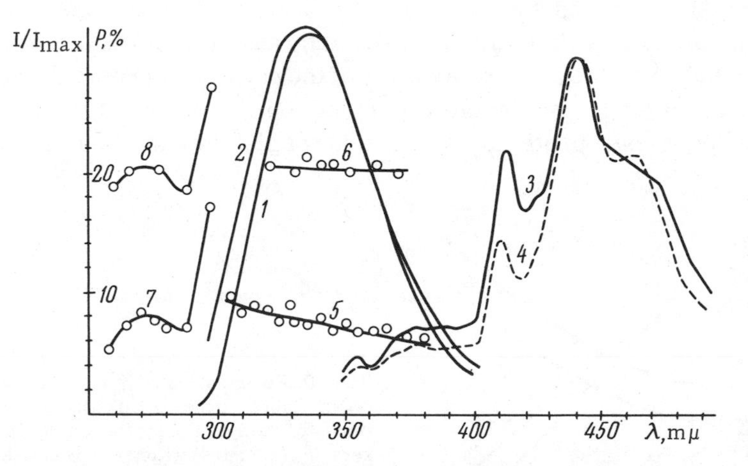

Fig. 39. Fluorescence spectra of globin (1) and hemoglobin (2), excited by λ = 265 mμ; phosphorescence spectra of globin (3) and hemoglobin (4) in 2 M LiI, t = -196°C, excited by λ = 265 mμ (room temperature, aqueous solutions); emission polarization spectra of fluorescence of globin (5) and hemoglobin (6) excited by λ = 265 mμ (room temperature); absorption polarization spectra of fluorescence of globin (7) and hemoglobin (8) excited by λ = 340 mμ (room temperature) (Konev, Bobrovich, and Chernitskii, 1965).

are twice as high above the x axis as that of globin, so that P_{265}^{350} of hemoglobin becomes 20%, i.e., acquires values characteristic of tryptophan in the free state. The fluorescence spectrum recorded through two monochromators and two polarizers in the same geometric conditions in which the polarization measurements are carried out shows that no extraneous luminescence is superimposed on the tryptophan luminescence.

Basically similar evidence in support of intertryptophan energy migration in proteins can be obtained by an investigation of the temperature dependence of fluorescence polarization. It is known (Levshin, 1951) that thermal quenching of the fluorescence of organic molecules is accompanied by a reduction of the lifetime of the fluorescent state. This in turn will reduce the probability of energy migration. In fact, Feofilov (1961) observed an increase in the polarization of the fluorescence of concentrated dye crystals with increase in temperature. In view of this the expected result of an increase in temperature to the plus side, when the possibility of reorientation of the tryptophan residues during the excited state is ruled out, would be an increase in the polarization accompanying the thermal quenching of proteins.

Two proteins, wool keratin and silk fibroin, were chosen as the main substances for this kind of investigation. This choice of substances is justified by the following factors. Firstly, both keratin and fibroin are solid samples, in which, at any reasonable temperature, the possibility of relaxation of the tryptophan residues during the excited state is completely ruled out. Secondly, the fluorescence of these proteins is most strongly depolarized, and, as a

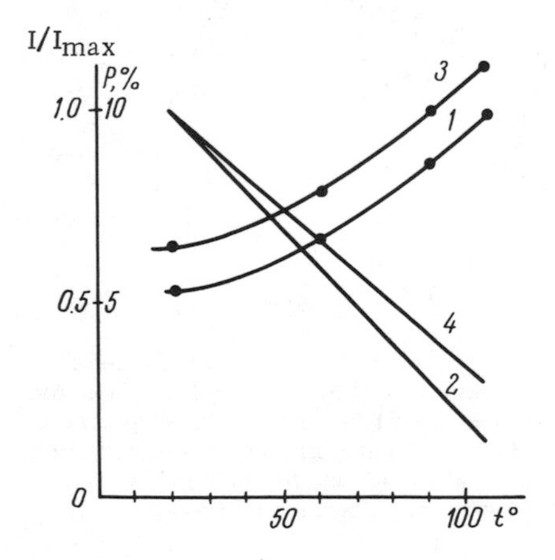

Fig. 40. Polarization P_{265}^{340} (1) and intensity (2) of fluorescence of silk fibroin as functions of temperature; the same for wool keratin (3 and 4) (Konev, Chernitskii, and Muraveinik, 1965).

result, the increase in fluorescence polarization with temperature increase will be all the more prominent.

Before the polarization measurements were made it was essential to ascertain that fluorescence quenching actually occurs in these substances and that it is quenching of the second kind.

The curves of $I_{fl} = f(T)$ for wool keratin and silk fibroin are given in Fig. 40. The figure shows that heating from 20 to 90°C leads to a steady decrease in the fluorescence intensity. At 90°C the fluorescence intensity is reduced by a factor of about 20. That the quenching is of a physical, and not a chemical, nature (for instance, thermal oxidation of tryptophan) is revealed by the reversibility of the quenching. When the temperature is lowered again, the fluorescence intensity takes its initial values at all points, including the initial point (20°C). This indicates the absence of chemical change in tryptophan in the investigated temperature range. Heating to higher temperatures (100 to 130°C) leads to hysteresis between the direct and reverse branch of temperature titration and even to visible yellowing of the samples. In other words, chemical destruction of tryptophan occurs at temperatures above 100°C. Hence, to avoid complicating the interpretation of the facts, we confined ourselves to an upper temperature threshold of 90°C.

The chemical identity of tryptophan molecules at 20 and 90°C can be demonstrated by recording the fluorescence spectra of proteins at these temperatures. As Figs. 41 and 42 show, heating leads only to some broadening of the spectra without any shift of their maxima. We can thus classify quenching of wool keratin and silk fibroin as quenching of the second order.

Figure 40 shows the results of measuring fluorescence polarization at different temperatures. With increase in temperature from 20 to 90°C, the fluorescence polarization gradually increases from 5 to 6% to 10 to 12%. The increase in the polarization, like the quenching of the fluorescence, is completely reversible.

The fluorescence polarization spectra of wool keratin (Fig. 41) and silk fibroin (Fig. 42) at 20 and 90°C have typical tryptophan shapes. They differ from one another at all points only in the absolute values of the polarization. The illustrated curves show that fluorescence depolarization develops in time, i.e., is of a migrational nature.

Finally, there is one other method of demonstrating intertryptophan energy migration in proteins. If we reduce the number of

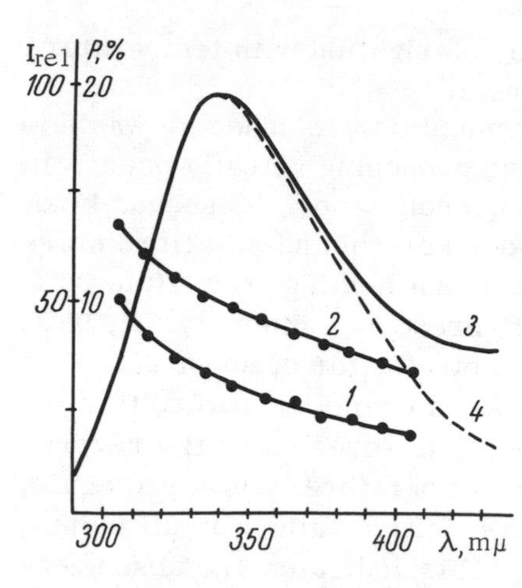

Fig. 41. Emission polarization spectrum of fluorescence (1) and fluorescence spectrum (4) of wool keratin at t = 20°C, λ_{exc} = 265 mμ; the same for 100°C (2 and 3) (Konev, Chernitskii, and Muraveinik, 1965).

tryptophan residues in the protein macromolecule in some way so that we finally obtain protein molecules containing on the average only one tryptophan residue, this will lead to the cessation of migration and, hence, to an increase in fluorescence polarization.

In fact, when the tryptophan residues in casein are gradually destroyed photochemically by irradiation of the protein solution with ultraviolet light (λ = 250 to 300 mμ), the fluorescence polarization increases from 10 to 18%. The experimental curve relating the polarization to the degree of destruction of the tryptophan residues $P = f(I_{fl,\ casein}) = f(t_{irrad})$ coincided with the curves calculated theoretically by the rule of addition of the polarizations

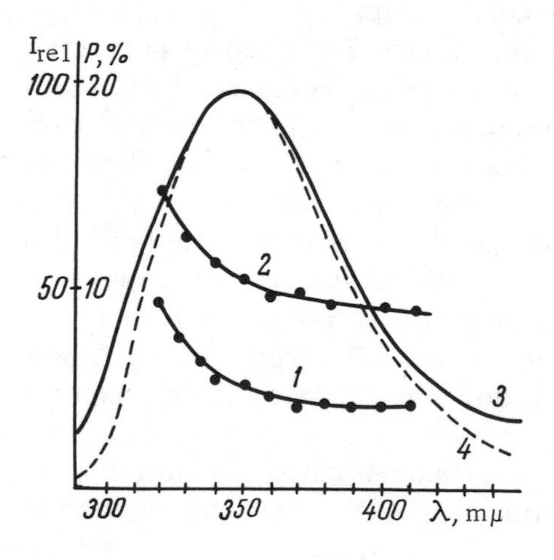

Fig. 42. Emission polarization spectrum of fluorescence (1) and fluorescence spectrum (4) of silk fibroin at t = 20°C, λ_{exc} = 265 mμ; the same for 100°C (2 and 3) (Konev, Chernitskii, and Muraveinik, 1965).

of the first two luminescent forms of casein molecules: casein with two tryptophan residues $\xrightarrow{h\nu}$ casein with one tryptophan residue $\xrightarrow{h\nu}$ casein with no tryptophan residue. We first calculated the concentrations of the first and second luminescent forms of the protein for any instant after the start of irradiation. The theoretical and experimental curves agreed in the case in which the polarization of the second form (tryptophan in protein, but no energy migration) was taken as 22%. In other words, the fluorescence polarization of tryptophan residues in protein when energy migration is excluded is close to that of tryptophan in the free state.

Thus, all the facts discussed above conclusively demonstrate the occurrence of energy migration in the protein macromolecule. This naturally gives rise to the question of how to reconcile these facts with those given by Weber, principally the fact of depolarization of the fluorescence of human serum albumin, which contains only one tryptophan residue. In this case there is no intertryptophan energy migration, but the fluorescence is still depolarized. This suggests that the mere incorporation of the tryptophan residue in the protein macromolecule without any migration is sufficient to polarize its fluorescence.

To analyze the mechanism of depolarization of the fluorescence of serum albumin, we point out first of all that Teale (1960 and 1961) and Weber (1960 and 1961) were too dogmatic in asserting the impossibility of energy migration from tyrosine to tryptophan.

The idea of the impossibility of energy migration from tyrosine to tryptophan is based on several facts. First of all, in the excitation spectrum of the fluorescence of serum albumin in the region of absorption of tyrosine residue, the tryptophan fluorescence is not sensitized but, on the contrary, is screened. Leaving aside the other evidence, we turn first of all to the fact that this screening is by no means complete, and it would probably be more correct to say that most of the tryptophan molecules act as a screen while the rest act as a sensitizer. This is indicated firstly by the experimental data which Teale himself obtained when he determined the fluorescence quantum yield of human serum albumin for different exciting wavelengths. For instance, at wavelength 300 mμ, where the absorption is due entirely to tryptophan, the quantum yield of the proteins is the same as tryptophan in the free state in aqueous solution (0.21). At 280 mμ, where 19.5% of the quanta are absorbed by tryptophan and 80.5% are absorbed by tyrosine, the quantum yield of the tryptophan fluorescence of albumin would fall to 0.04 in the case of pure screening. In fact, the fluorescence quantum yield of serum albumin at 280 mμ falls only to 0.07. In other words, tyrosine is both a screen and a sensitizer, and three out of every seven quanta emitted by tryptophan are due to energy migration from the tyrosine residues. Consequently, in the case of serum albumin, depolarization of the fluorescence of the single tryptophan residue can obviously be attributed to tyrosine-tryptophan energy migration. This explanation can be quantitatively tested (Bobrovich and Konev, 1965). On excitation by 300 mμ, where tyrosine does not absorb and tyrosine-tryptophan energy migration is absent, the polarization of the fluorescence of tryptophan and serum albumin will be the same. In fact, the polarization of the fluorescence of serum albumin and tryptophan are 32 and 35%, respectively (Fig. 32). Since the fluorescence due to energy transfer is depolarized, the depolarization due to energy migration from tyrosine at 280 mμ can be estimated

$$\frac{P}{P_0} = \frac{I_0}{\Sigma I_k}$$

where P is the fluorescence polarization in the case of energy migration, P_0 is the fluorescence polarization in the absence of energy migration, I_0 is the luminescence of molecules which emit without energy transfer, and ΣI_k is the total fluorescence intensity (Galanin, 1955).

As was explained above, for 280 mμ $I_0/\Sigma I_k$ is $4/7$. It is now easy to calculate the polarization of tryptophan fluorescence when tyrosine-tryptophan energy migration is absent. The polarization will be

$$P_0 = \frac{\Sigma I_k}{I_0} P = \frac{7}{4}\, 0.14 = 0.25$$

i.e., the same as in tryptophan itself in the free state for this wavelength. Hence, when a correction is introduced for the depolarization due to tyrosine-tryptophan energy migration, the fluorescence polarization spectra of the protein and free tryptophan become the same.

Thus, the oscillator nature of tryptophan residues incorporated in a protein macromolecule is not significantly altered, and the effect of the peptide bond or microenvironment cannot reduce the polarization, as is easily done by tryptophan-tryptophan energy migration. The example of human serum albumin now ceases to be a fact contradicting tryptophan-tryptophan and, on the contrary, becomes an indirect confirming fact. Weber's second point—the two inflections in the curve of the acid titration of the fluorescence of bovine serum albumin—does not contradict the idea of intertryptophan energy migration. The two inflections merely indicate that the rate constant of "neutral tryptophan"-"acid tryptophan" energy migration is less than that of the fluorescence pathway of deactivation of the tryptophan molecule.

Thus, all the facts discussed above lead to the conclusion that within the protein macromolecule there occur processes of migration of electronic excitation energy between tryptophan residues. As was shown earlier, tryptophan residues in protein can occur in three different electronic excited states with three corresponding electronic energy levels: ^1L_a and ^1L_b for the singlet excited states and ^3L_a for the triplet state. In this connection the first questions in the elucidation of the mechanisms of intertryptophan energy migration will be: Between what electronic levels of the excited molecule does energy transfer occur? Do all the electronic levels take part in energy migration and, if so, in what proportion?

Polarization measurements compel us to reject the implication of the singlet ^1L_b level and the triplet ^3L_a level in energy migration. This follows from the fact that molecules in the electronic excited ^1L_b state show hardly any fluorescence depolarization with increase in tryptophan concentration. As was mentioned earlier, the region of the fluorescence spectrum of tryptophan and proteins corresponding to the ^1L_b luminescence proper is not greatly depolarized with increase in tryptophan concentration in a solid solution of polyvinyl alcohol. Figure 18 illustrates this.

In an investigation of the fluorescence polarization spectra of indole in the region of concentration depolarization, Weber (1960 and 1961) came to exactly the opposite conclusion: Only the ^1L_b level is implicated in energy migration, and excitation of the ^1L_a band is ineffective for energy migration. It can be shown that this conclusion of Weber's not only does not follow from, but even contradicts, the experimental data. The data given by Weber himself for indole indicate first of all that, with increase in the concentration of the solution from 0.01 to 0.2 mole/liter, the fluorescence is depolarized in the excitation region of 260 to 265 mμ by a factor of 2.8, i.e., more than in the 290-mμ region (factor of 2.3). In fact, according to Weber's premise we should expect the reverse.

Migration would involve only ^1L_b, which absorbs at 290 mμ and, hence, would be at a maximum at this wavelength. This would lead to an increase, not a reduction, in the polarization, since in this case migration would depolarize the negative fluorescence and thus reduce the positive polarization of the total fluorescence.

The second experimental fact given in Weber's work is that, in concentrated solutions of phenol with an admixture of indole (1 M phenol + 0.01 M indole and 1 M phenol + 0.05 M indole) in the region of excitation of phenol (270 to 275 mμ), the polarization of the fluorescence of indole has small negative values, about 1 to 2%. Weber interprets this in the following way: Owing to the predominance of indirect excitation of indole by the energy migrating from phenol, and since only the negative oscillator is implicated in migration, negative polarization is found in the experiment. In fact, it is well known that even one act of energy migration between chaotically arranged molecules, irrespective of their oscillator nature, will lead to almost complete fluorescence depolarization (Galanin, 1956). Weber's data are more suggestive of zero polarization and certainly do not indicate negative polarization.

Finally, the third and apparently the only substantial argument is the absence in Weber's curves of concentration depolarization at 300 to 310 mμ. In fact, since only the ^1L_a oscillator in pure form is excited in this region, it would appear at first glance that this oscillator is inactive in energy migration. However, we showed in a previous chapter that fluorescence excited by 300 mμ or more is often not depolarized, because in this region there is considerable absorption by impurities, which, because their concentration is lower than that of the main form of tryptophan, naturally do not undergo depolarization. The fluorescence of spectrally pure proteins is depolarized at 300 mμ in the same way as at 265 mμ.

Thus, Weber's ideas regarding the inactivity of the ^1L_a oscillator in energy migration lack sound experimental foundation.

Despite the fact that triplet-triplet energy migration by the resonance exchange mechanism has been shown for many substances (Terenin and Ermolaev, 1954; and Ermolaev, 1964) and that even at room temperature proteins possess an appreciable steady concentration of molecules in the ^3L_a triplet state, we must also rule out the possibility of energy migration during the time spent in the phosphorescent triplet state. This conclusion follows inevitably from the equality of the migrational depolarization of the fluorescence and phosphorescence of chymotrypsinogen in comparison with free tryptophan. For this protein $P_{tryptophan}/P_{protein}$ is 2.8 for fluorescence and 2.7 for phosphorescence. The closeness of these ratios directly indicates that, during the time the tryptophan molecule is in the triplet state, there is no additional energy migration, which would be accompanied by additional depolarization. This is also indicated by the constancy of the polarization of the phosphorescence at different stages in its decay. The polarization of the phosphorescence at different phosphoroscope speeds (decay curve interval 10^{-1} to 10^{-3} sec) and when a mechanical shutter was used (decay curve interval 10^{-1} to 10 sec) was the same (Konev and Bobrovich, 1964). Hence, the entire energy transfer is effected during the lifetime of the ^1L_a singlet excited state. This is borne out well by the concentration depolarization of the tryptophan fluorescence in the region where the ^1L_a oscillator emits, as can be seen from the emission polarization spectra of the fluorescence (Konev, Chernitskii, and Bobrovich, 1964).

Thus, the ^1L_a electronic level is implicated in energy migration. By what physical mechanism is this energy transfer effected?

The absence of any requirement of structure factors and the possibility of energy migration between tryptophan molecules in solution and also in a polyvinyl alcohol film in both the native and denatured protein molecule exclude the participation of band conductivity or an exciton in energy transfer.

At the same time, all the main conditions for the occurrence of energy migration by the inductive resonance mechanism (Vladimirov and Konev, 1957) are fulfilled. Tryptophan exhibits luminescence, the fluorescence and absorption spectra of the molecules overlap, even though slightly, and the tryptophan residues in protein are separated by short distances of about 10 to 30 Å.

The mechanism of energy migration can also be determined from the nature of the relationship between the probability (rate) of energy migration and the distance between the energy donor and acceptor. The theory of energy migration by the inductive resonance mechanism, developed by Förster (1964 and 1959), is based, as is known, on the dipole nature of the interaction of the energy donor and acceptor. Consequently, the energy of the interaction between them is inversely proportional to the third power of the intermolecular distances. Since the probability of energy migration is directly proportional to the square of the interaction energy, its efficiency becomes inversely proportional to the sixth power of the intermolecular distances. According to Förster, the rate constant of energy migration is given by the following equation:

$$n_{S^* \to A^*} = \frac{9 \cdot 10^3 \ln 10 k^2}{128\pi^6 n^4 N \tau_s R^6} \int_0^\infty f_s(\nu)\, \varepsilon_A(\nu) \frac{d\nu}{\nu^4} \tag{1}$$

where $n_S{}^* \to A^*$ is the number of intermolecular transitions per second; ν is the wave number; $\epsilon_A(\nu)$ is the molecular extinction coefficient of the acceptor; $f_S(\nu)$ is the energy distribution in the fluorescence spectrum of the donor; N is Avogadro's number; τ_S is the lifetime of the excited state of the donor; n is the refractive index of the solvent; R is the mean distance between the donor and acceptor molecules; and k is an orientation factor, equal approximately to 0.666.

On introduction of the concept R_0, i.e., the critical distance at which the probabilities of energy migration and spontaneous deactivation (fluorescence) become equal, then

$$n_{S^* \to A^*} = \frac{1}{\tau_s} \left(\frac{R_0}{R} \right)^6 \tag{2}$$

where τ_S is the lifetime of the donor; $\tau_S = Q_S^0 \times \tau_S^0$ = quantum yield of the donor times its lifetime in the absence of energy migration.

The numerical value of R_0 for each donor-acceptor pair can be found from Eqs. (1) and (2):

$$R_0^6 = \frac{9 \cdot 10^3 \cdot \ln 10 k^2 \varphi_s^0}{128\pi^6 n^4 N \nu^4} \int_0^\infty f_s(\nu)\, \varepsilon_A(\nu)\, d\nu \tag{3}$$

Karreman and Steele (1957) introduced a simplified formula for calculation of the critical radius R_0:

$$R_0 = \sqrt[6]{0.95 \cdot 10^{-33} \frac{\tau j \tilde{\nu}}{\tilde{\nu}_0^2}} \tag{4}$$

where τ is the lifetime of the lowest singlet excited state of the donor; $j\tilde{\nu}$ is the integral (area) of overlap of the fluorescence spectrum of the energy donor and the absorption spectrum of the acceptor; and $\tilde{\nu}_0$ is the mean wave number between the fluorescence and absorption maxima of the donor.

Critical radii of energy migration between aromatic amino acids as calculated by Karreman, Steele, and Szent-Györgyi (1958) are shown in Table 13.

Table 13. Critical Radii of Energy Migration in Different Donor-Acceptor Pairs of Aromatic Amino Acids (Karreman, Steele, and Szent-Györgyi, 1958)

Donor-acceptor pairs	$\tau \cdot 10^8$	$\tilde{\nu}_0 \cdot 10^3$	$j\tilde{\nu} \cdot 10^{-8}$	R_0, Å
Phenylalanine-phenylalanine	1.10	37.1	0.0404	5.6
Phenylalanine-tyrosine	1.10	37.1	4.1000	12.0
Phenylalanine-tryptophan	1.10	37.1	21.1000	16.0
Tyrosine-tyrosine	0.91	34.4	0.458	8.3
Tyrosine-tryptophan	0.91	34.4	13.000	15.0
Tryptophan-tryptophan	0.20	36.6	0.326	6.3

It should be noted that Karreman et al. used the parameters of tryptophan in aqueous solutions for their calculations. In the case of tryptophan residues in most proteins, where the fluorescence spectrum is shifted 12 to 15 mμ on the average in the short-wave direction in comparison with aqueous solutions, the critical radius of intertryptophan energy migration will be much greater owing to the increase in the overlap integral $j\tilde{\nu}$. We carried out corresponding calculations of the critical radius for tryptophan residues in protein and in a polyvinyl alcohol film (Table 14). The values of the critical radius can also be found from the relationship between the fluorescence polarization and the intermolecular distances (concentration). According to Weber (1954 and 1960), the graph of the relationship between $1/P$ and the molar concentration C can be used to find the value $1/P_\infty$ for infinitely high dilution, and then from the formula

$$R_{\text{exp}} = \left[\frac{15S \cdot 10^3}{4\pi N (1/P_\infty + 1/3)} \right]^{1/6} (2a)^{1/2} \quad (5)$$

Table 14. The Critical Radius of Energy Migration between the Molecules of Tryptophan in Protein and Film of Polyvinyl Alcohol

Medium	$\tau \cdot 10^{-8}$	$\tilde{\nu}_0 \cdot 10^3$	$j\tilde{\nu} \cdot 10^{-8}$	R_0, Å
Polyvinyl alcohol (max.fl 320 mμ)	0.42	33.3	16.0	20.5
Tryptophan residue in chymotrypsinogen (max. fl 331 mμ)	0.16	32.8	10.0	16.0
Tryptophan residue in pepsin (max. fl 342 mμ)	0.45	32.2	4.0	17.0

(where N is Avogadro's number, a is the mean radius of the molecule, and S is the slope of the curve), the critical radius is determined.

We suggest a simpler graphical method of polarization determination of the critical radius. This method gives results which agree satisfactorily with those obtained by Weber's method. Since at the point R_0 the probabilities of emission and migration are equal, about half of the quanta are emitted in the primary excitation and the other half are emitted after energy migration. This statement is based on the fact that (as will be shown a little further on) the fluorescence quantum yields of tryptophan molecules for primary excitation and excitation due to energy migration are equal.

Assuming this second luminescence to be completely depolarized, it is easy to see from the rule of addition of polarizations

$$P = \frac{\Sigma p_i I_i}{\Sigma I_i} = \frac{0.5\, I p_0}{0.5 I + 0.5 I} + \frac{0.5 I \cdot 0}{0.5 I + 0.5 I} = \frac{p_0}{2}$$

that, when the intermolecular distances R_0 are equal, the fluorescence polarization of the system will be half of the maximum observed in highly diluted solutions without energy migration. The concentration of the substance, and hence R_0, is easily found graphically on the curve of the relationship p = f (c). If the values of R_0 found from polarization measurements and calculated theoretically are compared, they are found to be fairly close to one another. For instance, according to Weber's calculation (1961) for indole in propylene glycol at -70°C, the experimental value is 17 Å, and the theoretical, 23 Å (apparently a slight overestimate); for tryptophan in polyvinyl alcohol the theoretical value is 20.5 Å, and the experimental value is 21.5 Å (Konev, 1964). The agreement between the theoretical and experimental values supports the resonance mechanism of energy migration between molecules.

As regards tryptophan in protein, its residues in the case of a mean statistical distribution will usually be separated from one another by distances of 20 Å, i.e., by distances very close to the critical radius. The same agreement between the theoretical and experimental values of R_0 is observed for tyrosine (phenol): R_0 = 17 Å is the experimental value, and 11 Å, the theoretical value (Weber, 1961).

Of very great importance for the intramolecular energetics of the protein macromolecule is the question of the quantum yield of intertryptophan energy migration. Does the energy transfer process involve loss by dissipation as heat, or does the process have a 100% quantum yield? In other words, does concentration quenching of tryptophan fluorescence occur?

Concentration quenching of fluorescence is well known in the case of dyes. There are two mutually exclusive views on the causes of concentration quenching. The first of them, expressed by Levshin as far back as 1927, is that energy migration in itself does not entail loss of electronic excitation energy and that the quenching effect is due, firstly, to inactive absorption by associates appearing at high concentrations and, secondly, to energy migration to them from the monomers. The other viewpoint regarding quenching at the moment of migration from monomer is due to Galanin (1960) and is not nearly so well substantiated.

Concentration quenching of the fluorescence of aqueous solutions of tryptophan was described by Vladimirov in 1957 and was confirmed by Konev, Katibnikov, and Petrova in 1961 for a water-glycerol solution. However, in contrast to dyes, no dimers or associates are formed in concentrated aqueous solutions of tryptophan. This is indicated by the way in which concentration quenching depends on the wavelength of the exciting light. Vladimirov (1957) found that the shapes of the concentration-quenching curves were the same for three exciting wavelengths, and Konev, Katibnikov, and Lyskova (1964) obtained the same result for the whole 220- to 300-mμ region (Fig. 43). This indicates the absence of associates in the investigated solutions, since otherwise there would be more pronounced fluorescence quenching for the wavelengths which coincide with the absorption maximum of the associate. The similar quenching for the entire 220- to 300-mμ region indicates the constancy of the fluorescence excitation spectrum, i.e., the constancy of the absorption spectrum of the whole pool of molecules in the solution with increase in concentration. Rejecting the possibility of

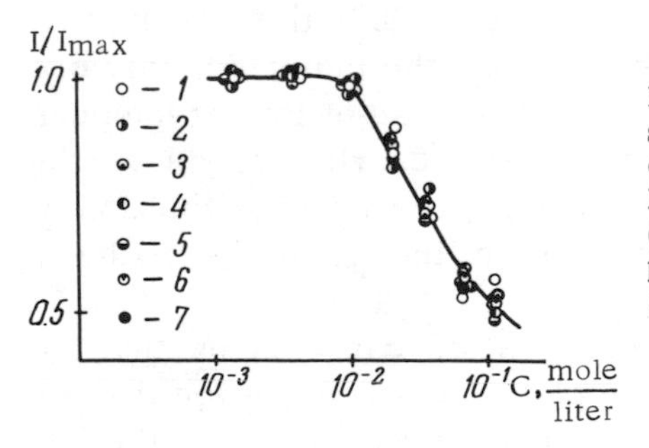

Fig. 43. Concentration quenching of aqueous solutions of tryptophan (pH 10) on excitation of fluorescence by light of different wavelengths: (1) 220, (2) 248, (3) 254, (4) 265, (5) 280, (6) 290, (7) 302 mμ, at room temperature (Konev, Katibnikov, and Lyskova, 1964).

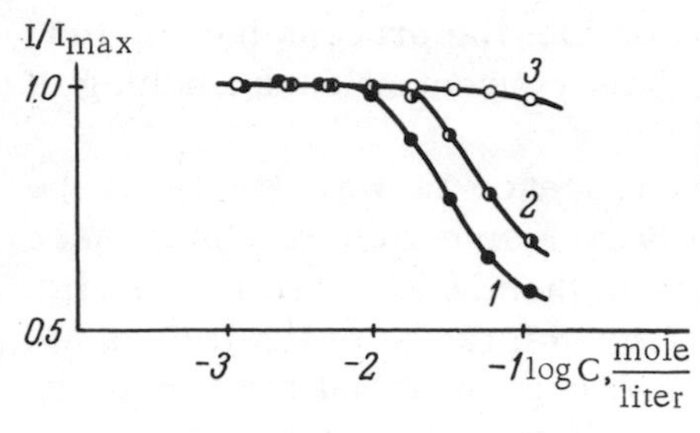

Fig. 44. Concentration quenching of fluorescence of tryptophan solutions: (1) in aqueous solution (pH 10); (2) in water-glycerol mixture (1:1, pH 10); (3) in dehydrated glycerol (dry NaOH was added until pH was 10 according to indicator paper); at room temperature (Konev, Katibnikov, and Lyskova, 1964).

concentration quenching due to the formation of dimers or other associates, we must assume that the quenching is due to energy migration from monomer to monomer. Yet such a hypothesis has been proved false. It was found (Konev, Katibnikov, and Lyskova, 1964) that with increase in viscosity of the solvent the concentration quenching gradually decreases and at high viscosities (1000 cp or more) disappears altogether (Fig. 44). In the case of solid films of polyvinyl alcohol, concentration quenching could not be observed even at concentrations as high as 1 mole/liter, where tryptophan fluorescence is completely depolarized due to energy migration (Fig. 14a, curve 1). Hence, in contrast to dyes, where quenching is often of a static nature and does not depend on the viscosity (Levshin and Krotova, 1960), the quenching of tryptophan fluorescence is of a pronounced kinetic nature and is not due to energy migration processes. Conditions for efficient quenching are created only when an excited tryptophan molecule collides with an unexcited molecule. The most important fact at the moment is that energy migration does not lead to a reduction of the fluorescence quantum yield. We can thus conclude that the quantum yield of the migration process between tryptophan molecules is 1.0. Moreover, excitation of molecules by energy migration is approximately five times more efficient than their excitation by light quanta. Since the quantum yield of tryptophan fluorescence is 0.2, this means that out of five exciting light quanta absorbed by the molecule, only one gives rise to a fluorescent molecule. In the case of intertryptophan energy migration, each energy quantum absorbed produces the fluorescent state. Hence, the migration pathway of excitation of molecules is much more efficient than the radiative pathway (direct excitation by quanta).

Taking the above into consideration, we can assume that the characteristic feature of intertryptophan energy migration in proteins, effected by the inductive resonance mechanism via the ^1L_a

singlet levels of the tryptophan molecules, is the 100% quantum yield of energy transfer.

However, in the assessment of the specific role of energy migration in the energetics of the protein macromolecule, a feature of special importance, in addition to the quantum yield of energy migration, is the efficiency, i.e., the percentage of "delocalized" fluorescent excited states of tryptophan (those which have changed the site of their original appearance) in relation to the total number of fluorescent states produced in a unit of time. The probability that the fluorescence state of the tryptophan residue in the protein macromolecule due to absorption of a light quantum will be transferred to another tryptophan residue in another part of the protein molecule decides the role of energy migration in photochemistry and photobiology.

Agreeing with Galanin that the fluorescence, even after one act of energy transfer is almost completely depolarized, we write the previously given formula in the following form:

$$\frac{P_0 - P}{P_0} = \frac{\Sigma I_k - I_0}{\Sigma I_k} = \eta$$

where η gives the efficiency of energy migration.

Since the rate constant of energy migration, and hence its efficiency, is proportional to

$$E = \frac{1}{(R_0/R)^6 + 1}$$

the efficiency of energy migration will depend greatly on the distances between the tryptophan residues, i.e., on the intraprotein topography of this amino acid. It follows from this formula, for instance, that, for R_0 = 16 Å for tryptophan residues in protein, a reduction of the distance by only 3 Å will be accompanied by an increase in the efficiency of energy migration from 50 to 77% and that a reduction of the distance to 10 Å will lead to an increase in the efficiency of energy migration to 95%. Consequently, a priori one can expect considerable variations in the efficiency of energy migration in different proteins. In fact, an assessment from the formula cited above gives the following values of efficiency of energy migration (Table 15).

In the case of wool keratin the polarization of fluorescence is closely dependent on the conformation of the protein molecule, i.e., on the configuration of the polypeptide chain (Konev and Katibnikov, 1961). Conversion of the α configuration of the polypeptide chain

Table 15. Efficiency of Intertryptophan Energy Migration in Various Proteins

Protein	Polarization P_{270}^{335}	Efficiency of energy migration, %
Human serum albumin	14.0	0*
Pepsin	11.8	55
Edestin	8.6	67
Squash-seed gamma globulin	15.5	40
Chymotrypsinogen	9.5	64
Trypsin	12.2	53
Chymotrypsin	12.0	54
Globin	12.2	53
Hemoglobin	21.0	19
Silk fibroin	6.0	77
Wool keratin	5.0	81
Cytochrome c	7.0	73

*The fluorescence depolarization is due to tyrosine-tryptophan energy migration.

to the β configuration by stretching the wool fiber in steam to double its length leads to an increase in the fluorescence polarization from 4% to 20 to 25% (Figs. 18 and 33). The data of X-ray analysis show that conversion from the α to the β configuration is reversible.

In complete correspondence with this, after removal of the stretching force, the fluorescence polarization is gradually reduced to its initial level (4 to 5%). Hence, in this case changes in the fluorescence polarization reflect the structural changes occurring in protein molecules. In this case, the changes in the fluorescence polarization are due mainly to orientation, and not migration, effects. The orientational nature of the polarized luminescence of wool keratin is indicated by several facts, primarily the depolarization of the fluorescence of wool stretched in steam when its long axis is oriented perpendicular to the vector of the exciting light or when the fibers are randomly oriented. This directly indicates that intense energy migration processes continue to occur in the majority of the tryptophan residues and that only some residues, owing to their orientation, emit polarized luminescence.

Secondly, the presence of oriented regions of tryptophan residues is indicated by the occurrence of spontaneous polarization of fluorescence excited by natural unpolarized light (+4 and -4%).

Finally, the orientation effect is manifested in the possibility of obtaining an absorption polarization spectrum of fluorescence on excitation of vertically oriented steam-stretched wool by exciting light with its electric vector horizontally oriented (Fig. 33, curve 8). This polarization spectrum, as pointed out above, reflects the absorption spectrum of the ^1L_b electronic transition of oriented tryptophan residues and can occur only when the molecular axes are oriented to the axis of the wool fiber (Fig. 45). The main biologically significant result of energy migration is the displacement of the electronic excited states in space, which is like an increase in the cross section for the capture of ultraviolet quanta. Another important role of the primary photophysical processes following the act of energy absorption may consist in the temporal separation of the moment of absorption of the quantum and the accomplishment of its chemical action. This increase in the times of action of quanta and the enlargement of the spatial sphere of their action can be a very important factor in photo- and radiobiology. This gives rise to another question closely linked with energy migration: the possibility of long conservation of electronic excitation energy in protein systems. Since vital processes take place at positive temperatures, we are interested primarily in the question of the possible long conservation of electronic excitation energy at relatively high, i.e., room temperatures. At the same time, we shall try to clarify the role of tryptophan in energy conservation mechanisms.

The first reports of the possibility of afterglow of dry protein powders at room temperature are found in the work of Vladimirov

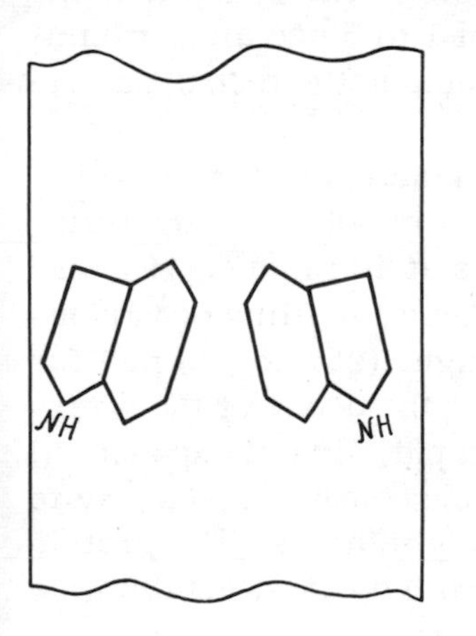

Fig. 45. Arrangement of tryptophan residues relative to keratin fibril.

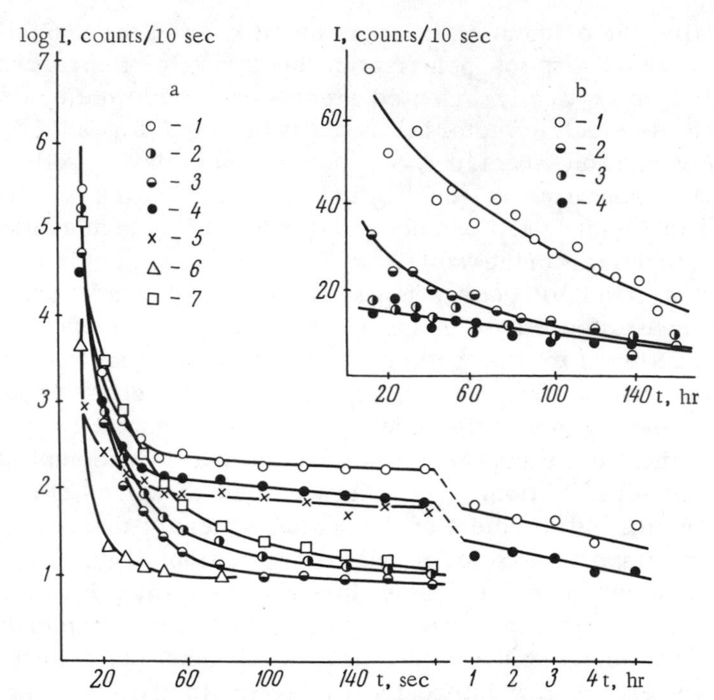

Fig. 46. Kinetic curves of decay of intensity of afterglow of various proteins in (a) film after exposure to an SVD-120A lamp at a distance of 10 cm for 15 sec: (1) wool keratin, (2) amylase in film, (3) human serum albumin in film, (4) silk fibroin, (5) barley radicles, (6) clupein, (7) pepsin; (b) aqueous solutions: (1) human serum gamma globulin, (2) human serum albumin, (3) clupein, (4) asparagine (0.1% aqueous solution) (Konev and Katibnikov, 1961).

and Litvin (1960), who recorded an afterglow with sharp maxima at 435 and 455 to 460 mμ during a period of 1 to 5 sec after illumination with ultraviolet light. Yet they could not detect the afterglow of proteins in solution.

Konev and Katibnikov investigated the kinetics of the afterglow of proteins in solution and in transparent films obtained by drying concentrated aqueous solutions of proteins at 40 to 50°C (1961 and 1962). Figure 46 shows the afterglow curves for films of various proteins after irradiation by ultraviolet light (250 to 400 mμ) for 15 sec. A study of these curves shows that the decay of the afterglow in protein films has two phases: a rapid, almost exponential reduction of intensity in the first 30 sec subsequently gives way to a slowly decaying luminescence lasting many minutes. The protein clupein, which does not contain aromatic amino acids, exhibits, as measurements showed, an afterglow with an intensity 1.5 to 2 orders less than that of the afterglow of tryptophan-containing proteins. This fact indicates a causal relationship between the tryptophan content of a protein and its ability to exhibit an afterglow at room temperature. The first, almost exponential part of the afterglow of proteins is possibly the β phosphorescence of protein tryptophan at room temperature. That this part of the afterglow is due to tryptophan molecules is clearly revealed by the shape of the excitation spectra of the first phase of the afterglow, which show the two absorption bands of tryptophan residues in protein, at 280 and 225 mμ, separated by a minimum at 240 mμ (Fig. 47). The afterglow spectrum as a whole coincides with the phosphorescence spectrum of proteins at low temperatures, but it is shifted toward the long-wave side owing to the fact that the luminescence maximum is shifted from the vibrational band at 438 mμ to the shoulder at 470 to 480 mμ. The gradual redistribution of the maximal intensities, expressed in an increase of the intensities of the long-wave components, is seen when the spectra are recorded as the temperature increases from -(30 to 50)°C to room temperature. In several proteins, such as milk casein granules, the structure of the phosphorescence spectrum remains the same even at room temperature. On the assumption of the tryptophan nature of the afterglow of proteins at room temperature, there remains the question of the

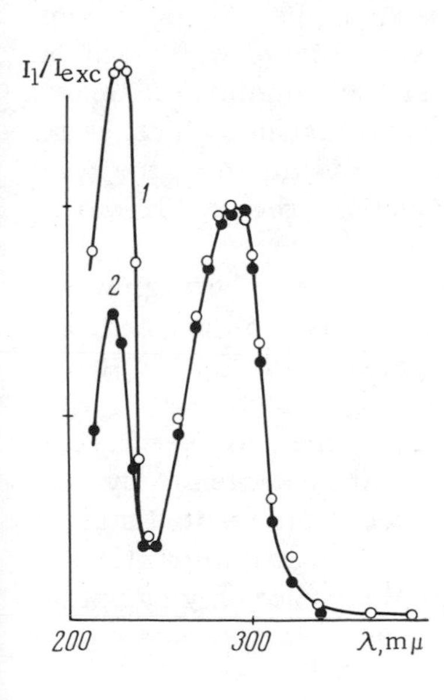

Fig. 47. Excitation spectra of the first stage of afterglow of pepsin (1) and amylase (2) films at room temperature (Katibnikov and Konev, 1962).

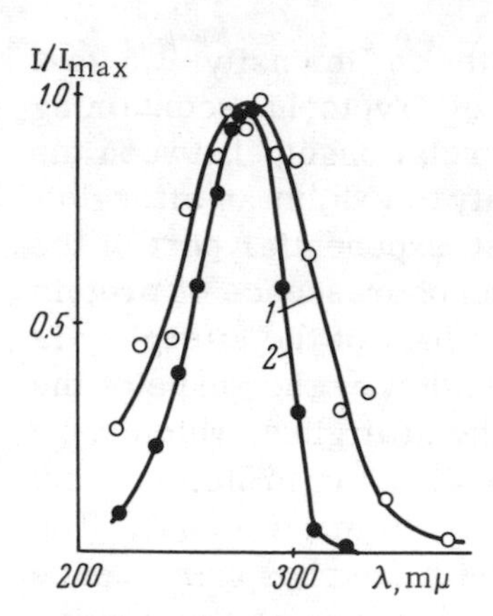

Fig. 48. Excitation spectra of afterglow of Na-carboxymethylcellulose (1) and clupein (2) films at room temperature (Katibnikov and Konev, 1962).

centers responsible for the quite distinct, though 1.5 to 2 orders weaker, afterglow of clupein, which does not contain any aromatic amino acids.

Figure 48 shows that the excitation spectrum of the clupein afterglow differs distinctly from the corresponding spectra for tryptophan-containing proteins. The clupein spectrum contains only one band, with a maximum at 270 mμ. Light with wavelengths greater than 300 mμ and less than 225 mμ does not produce an afterglow. Another interesting fact is that the excitation spectrum for the clupein afterglow does not coincide at all with its absorption spectrum, which has no absorption band in the 270-mμ region. It is characteristic that the excitation spectrum of water-soluble Na-carboxymethylcellulose in a film gives the same maximum with the same intensity spectrum. It is possible that in both cases the chromophores are carboxyl groups, the maximum of the weak absorption of which lies in the 270- to 280-mμ region (Bredereck et al., 1953; and Schur, 1957).

Hence the centers responsible for the clupein afterglow are different from those in tryptophan-containing proteins and, accordingly, the afterglow intensity is approximately 0.01 of that of other proteins.

Thus, a special feature of the energetics of the protein molecule is the possibility of the creation of very appreciable steady concentrations of molecules in the triplet electronic excited state in them at low temperature. It is characteristic that triplet states of tryptophan residues in proteins can be produced not only by optical excitation, but also by chemical excitation. The second kinetic stage of the afterglow of proteins is of special interest. The most intense afterglow at this stage is observed in proteins which are rich in sulfur-containing amino acids—wool keratin and pepsin. The different mechanism of the first and second stages of the luminescence is revealed not only in the kinetics, but also in the obvious thermally activated nature of the second stage. An increase in temperature to 75°C leads to a pronounced (tenfold) reduction of the emission of wool keratin in the first stage and, on the other hand, to an increase in the second stage (Fig. 49). The increase in afterglow intensity with increase in temperature for the second stage conforms to an exponential law (Fig. 50). The activation energy for wool keratin is 21.5 kcal/mole. It is remarkable that Riehl (1956) found the same value of activation energy for the electrical conduction of water-containing proteins. An activational

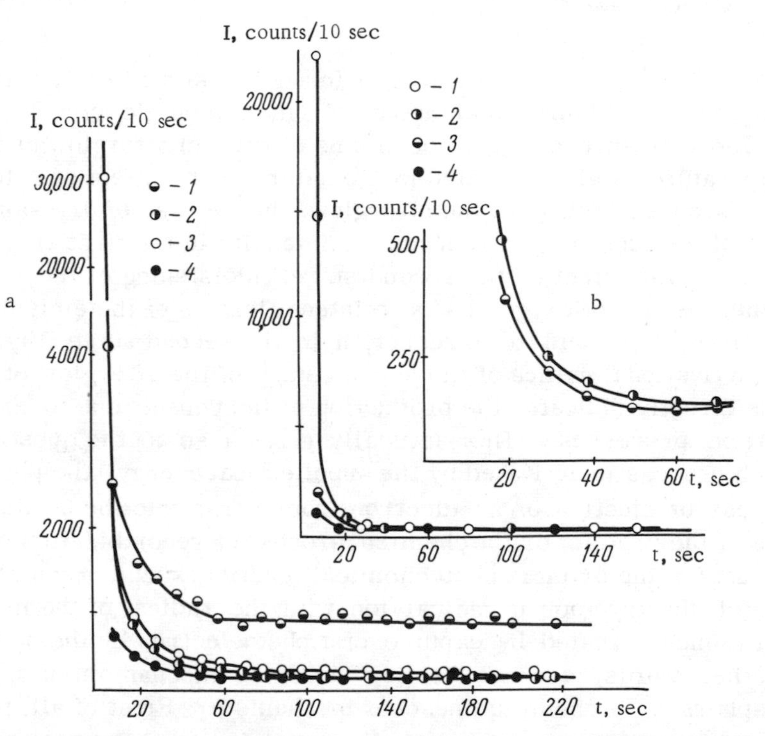

Fig. 49. (a) Effect of temperature on kinetics of afterglow of wool after irradiation with ultraviolet light (250 to 380 mμ) for 15 sec: (1) wool at 75°C, (2) damp wool at 20°C, (3) wool at 20°C, (4) damp wool at 75°C; (b) effect of oxygen on kinetics of afterglow of wool after irradiation in the same conditions: (1) damp wool in vacuum, (2) wool in vacuum, (3) wool in air, and (4) damp wool in air (Konev and Katibnikov, 1961).

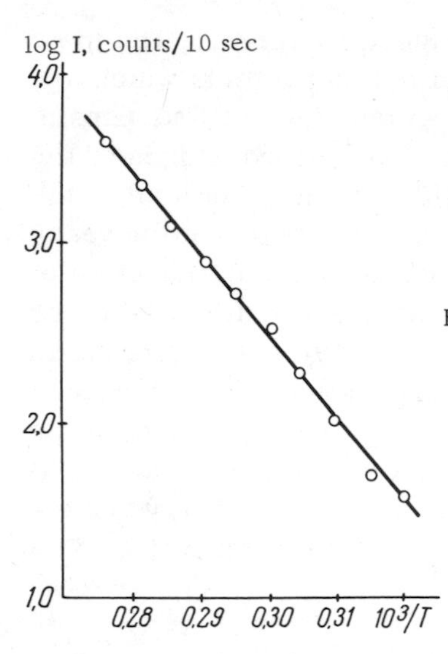

Fig. 50. Intensity of afterglow of wool as a function of temperature (Konev and Katibnikov, 1961).

type of afterglow has also been found for silk fibroin, amylase films, pepsin, albumin, and human serum gamma globulin.

The existence of two mechanisms of protein afterglow at room temperature is also revealed in the nature of the effect of external factors on each stage of the afterglow. In the case of dry samples of wool keratin, oxygen reduces the intensity in the first stage and has no great effect on the second stage. Moistening of the sample in the absence of oxygen leads to intensification of the emission in the first stage and to a reduction in the second stage (Fig. 49). The activational nature of the second stage of the afterglow of proteins directly indicates the production of light quanta due to recombination processes. This naturally gives rise to the question of which process is activated by the supplied heat energy: the physical process of ejection of an electron from a trap into the conduction band of the protein, or the chemical process of recombination of the products of the primary photochemical reaction, such as recombination of the tryptophan radical ion with the radical of the microenvironment created by capture of a photoelectron by the solvent. In other words, are we dealing with the mechanism of crystal phosphors or a chemiluminescent mechanism? First of all, paramagnetic resonance data show that proteins exposed to ionizing or ultraviolet radiation give a signal due to unpaired electrons (Blyumenfel'd, 1957; Éidus and Kayushin, 1960; Éidus, 1960; Patten and Gordy, 1960; and others). Of most interest in this case are the data of Sapezhinskii and Silaev (1965) and Sapezhinskii and Émanuel (1965), who observed good parallelism between the reduction of the EPR signal and the afterglow intensity in protein systems exposed to ultraviolet light. In view of this information it is certain that some atoms with unpaired electrons are implicated in the afterglow mechanism. However, the carriers of the EPR signal may be either products of the photochemical reaction (radicals) or electrons captured in traps close to the conduction band. A fact which supports the semiconducting mechanism is the equality of the activation energies for the electrical conduction and afterglow of proteins. However, the presence of the second stage in the afterglow of native and denatured proteins rules in favor of the chemiluminescent mechanism of afterglow induction. This mechanism of induction of the second stage of the afterglow is made all the more probable by the fact that a distinct chemiluminescence, dependent on oxygen pressure, is found in the case of human gamma globulin in aqueous solution (Konev and Katibnikov, 1961) and for amino acids themselves (Vladimirov, Litvin, and Li Man-ch'i, 1962). This also explains the very pronounced dependence of the second stage of the afterglow on the viscosity and rigidity of organization of the protein structure. The viscosity of the medium is apparently the factor which prevents the instantaneous recombination of the free radicals formed. Further investigation of the features of this stage of the afterglow suggests the following scheme of processes:

1. tryptophan + $h\nu \rightarrow$ tryptophan ^{S*}
2. tryptophan $^{S*} \rightarrow$ tryptophan ^{T*}
3. tryptophan ^{T*} + A $\rightleftarrows$ tryptophan $^{R+}$ + A - $\bar{e}$
4. tryptophan $^{R+}$ + $\bar{e}$ - A $\rightleftarrows$ tryptophan^{T*} + A
5. tryptophan $^{T*} \rightarrow$ tryptophan + $h\nu_{ph}$

The role of tryptophan as an acceptor of ultraviolet quanta is confirmed by the coincidence of the excitation spectra of the first and second stages of the afterglow (experiments with light filters). The first stage exhibits a tryptophan excitation spectrum on monochromatic excitation.

The appearance of light quanta in the second stage as a consequence of a triplet-singlet transition in the recombined tryptophan molecule is revealed by an investigation of the spectral composition of this luminescence. The complete absence of luminescence with a maximum at 350 mμ, corresponding to a singlet-singlet transition, and, on the other hand, the similarity of the spectral composition of the second stage of the afterglow of tryptophan phosphorescence (although slightly shifted in the long-wave direction) confirm the above scheme. The proposed scheme of induction of the long afterglow of tryptophan residues is not in any way unusual, generally speaking. The long afterglow of several organic substances in glasses, due to dissociation of a photoelectron and its retention in the unpaired state in the solvent, has been observed by Lewis and Kasha (1944), Kasha (1947), Linschitz et al. (1954), and several other authors.

The above scheme has a certain measure of specificity in that in this case the process proceeds via triplet electronic excited states of tryptophan and not via singlet states. We are dealing in effect with a specific type of chemiphosphorescence, the possibility of which has been questioned by many authors, including such a well-known investigator as Reid (1960).

Hence, in the most general form the chain of events leading to the afterglow can be described as follows:

1. Absorption of energy by tryptophan molecules with their conversion to the triplet state and dissociation of photoelectrons.
2. Capture and retention of photoelectrons by the supermolecular environment and the formation of free radicals.
3. Thermally activated reverse reaction of recombination of the tryptophan residue with the dissociated electrons and emission from the tryptophan triplet levels.

Chapter 4

LUMINESCENCE OF NUCLEIC ACIDS AND ENERGY MIGRATION IN THEM

The history of research on the luminescence of nucleic acids is very much like the history of research on protein luminescence. In the thirties and forties there appeared a considerable number of papers which described the visible (violet, blue, purple, green) fluorescence of free bases and their derivatives and of nucleic acids themselves, due to excitation by light of wavelength 365 mμ (Neyroth and Loofbourow, 1931; Euler, Brandt, and Neumüller, 1935; Loofbourow, Cook, and Stimson, 1938; Stimson and Reater, 1941; and others). For instance, in a recent investigation the luminescence of 20 derivatives of nitrogenous bases, including uracil, 2,6-dichloropyrimidine, cytosine, isocytosine, guanine, isoguanine, adenosine phosphoric acid, yeast, and calf-thymus nucleic acids in the solid state, in NaOH, NH_4OH, and H_2SO_4, was detected visually. The reduction of luminescence intensity on purification, the possibility of exciting it outside the absorption band, and the lowest luminescence intensity of guanine in an acid medium, i.e., where it exhibits greater fluorescence intensity than other bases, leave no doubt that the above-cited authors were dealing with the luminescence of impurities and not of the main substance.

Subsequent, more refined spectrofluorometric investigations showed that neither purine nor pyrimidine bases in neutral aqueous solutions at room temperature are capable of luminescence either in the ultraviolet or visible regions.

In the case of solutions at room temperature only the fluorescence of purine bases—guanine and adenine—in a highly acid medium (0.1 N HCl) has been described (Duggan et al., 1957; Barskii, 1959; and Agroskin, Korolev, et al., 1961). On the curve of acid titration of the fluorescence intensity of guanine I = f (pH), which is S-shaped, the half-quenching point is close to the point of conversion of half of the guanine molecules to the tautomeric form (Barenboim, 1963). In an acid medium the fluorescence maximum of guanine lies at 355 mμ and that of adenine at 365 mμ. The fluorescence excitation spectra are similar to, though not identical with, their absorption spectra (Agroskin, Korolev, et al., 1961).

In 7 N hydrochloric acid the fluorescence of guanine has a maximum at 365 mμ (Duggan et al., 1957; and Barskii, 1959), and at pH 1 the maximum is shifted in the short-wave direction to 352 mμ (Barskii, 1959).

The phosphorescence of adenine compounds at 194° K was described by Steele and Szent-Györgyi (1957). The time course of the decay was satisfactorily described by a first-order equation, and the rate constant of the decay did not depend on the temperature. In view of this the authors attributed the afterglow of adenine compounds to the triplet state of the molecules. In contrast to Steele and Szent-Györgyi, who investigated frozen, snow-like specimens, Bersohn and Isenberg (1964) worked with transparent water-glycerol glasses at 77°K, and using a phosphoroscope with a time resolution of 10^{-3} sec they recorded a well-structured phosphorescence spectrum with maxima at 390, 415, 445, and 460 mμ for adenine (deoxyadenosine monophosphate) and a structureless spectrum with a maximum at 420 mμ for guanine (deoxyguanosine monophosphate). Derivatives of pyrimidine bases in the same conditions did not exhibit detectable fluorescence. We can surmise that freezing rigidly fixes the ring of bases and obstructs the extremely efficient radiationless pathway of energy dissipation. In a way similar to freezing, other methods of immobilizing the skeleton of the molecule lead to the appearance of luminescent ability. Fluorescence appears when bases are absorbed on chromatographic paper (Barskii, 1959) and when dry powders of bases and their derivatives are frozen to liquid-oxygen temperature (90°K). In the latter case distinct luminescence appears, but it has a very low yield (apparently about 1%), with smeared maxima lying in the range of 320 to 400 mμ. The positions of the maxima are shown in Table 16, which is taken from the paper of Agroskin, Korolev, et al. (1960). Apparently the solid or, more precisely, powder aggregate state of the specimen is not at all essential for

Table 16. Positions of Luminescence Maxima of Powdered Nucleic Acids and Their Chromophores at t = 90°K (Agroskin, Korolev, et al., 1960)

Substance	Spectral maximum, mμ		Substance	Spectral maximum, mμ	
	λ_{exc} 260–280 mμ	λ_{exc} 296–313 mμ		λ_{exc} 260–280 mμ	λ_{exc} 296–313 mμ
Adenine	355	355	Uracil	355	430
Adenosine	375	375	Uridine-5-monophosphate	320	430
Adenosine-5-monophosphate	360	350	Uridine-5-triphosphate	330	340
Guanine	370	370	Thymine	310	395
Guanosine	325	355	Thymidine-5-monophosphate	335	375
Guanosine-5-monophosphate	370	360	RNA	350–360	380–400
Cytosine	400	400	DNA	340–360	350–380
Cytidine-5-monophosphate	330	340	Various polyphosphates	320–350	320–350

the appearance of fluorescence. Bersohn and Isenberg (1964) recorded the fluorescence of thymine and cytosine in water-glycerol crystals at 77° K with the same maximum at 340 to 350 mμ, which is slightly different from the values found by Soviet authors owing to the inaccurate correction for the spectral sensitivity in the latter case.

The first attempts to detect ultraviolet luminescence of nucleic acids were unsuccessful, mainly because of the low sensitivity of the apparatus (Shore and Pardee, 1956).

The ability of nucleic acids to fluoresce and also to phosphoresce was demonstrated later. Konev (1957) described the fluorescence of calf-thymus nucleohistone, and Douzou et al. (1961) described the fluorescence and phosphorescence of the same substance.

For calf-thymus DNA, salmon-sperm DNA, and T2-phage DNA, Isenberg et al. (1964) obtained phosphorescence excitation spectra with a maximum at 295 to 300 mμ, which did not coincide with the main long-wave absorption maximum. The French authors obtained similar excitation spectra for DNA (300 mμ), polyuridylic acid (311 mμ) and polyadenylic acid (270 mμ). Accepting Kasha's interpretation (1961) of the nature of the electronic absorption spectra of nitrogenous bases, the French authors postulated that the fluorescence of nucleic acids was due entirely to $n-\pi$ transitions.

According to the data of these authors, the DNA fluorescence spectrum has a main maximum at 355 mμ and two weaker maxima at 325 and 390 mμ. It should be pointed out, however, that the quantum yield of the fluorescence of DNA in aqueous solution is very low. Distinct luminescence of nucleic acids in aqueous solutions at room temperature is observed only in a highly acid medium (Agroskin, Korolev, et al., 1961; Barenboim, 1963; and Douzou et al., 1962). The fluorescence maximum of DNA lies at 355 mμ and coincides with the fluorescence maximum of guanine. The excitation spectrum also coincides with the absorption spectrum of guanine. For low-polymer and high-polymer yeast RNA and low-polymer and high-polymer calf-thymus DNA, the excitation spectrum consists of maxima of various heights and shapes at 250 and 275 to 280 mμ. The noncoincidence of this spectrum with the absorption spectrum of the RNA, and also of the DNA, macromolecule as a whole indicates not only the predominance of guanine luminescence but also the absence of adequately efficient migration to it from the other bases. The exceptional role of guanine is confirmed also by the fact that purine-free nucleic acid (apurinic acid) does not exhibit luminescence, whereas the removal of pyrimidines has little effect on the spectral luminescent properties (Agroskin, Korolev, et al., 1961). According to Barenboim's data (1963), DNA luminesces in solution only from pH 4.2, and the S-shaped curve of titration of fluorescence intensity coincides with that of acid titration of guanine. This provides further evidence in support of the conclusion that fluorescence of DNA is due to the acid tautomeric form of guanine. But even in these conditions its quantum yield is very low, not more than 0.3 to 0.5% (Konev, 1964).

The most interesting feature of the luminescence of nucleic acids in aqueous solutions at room temperature is probably the fact that, on transition to weakly alkaline reactions in DNA (phosphate buffer pH 7.2) (Agroskin, Korolev, et al., 1961) and in the case of nucleohistone for aqueous and aqueous-phosphate solvents (Konev, 1957), a fairly intense fluorescence appears again, but the main excitation maximum is at 290 to 300 mμ. There are some grounds for ascribing this luminescence to n-π transitions in the DNA macromolecule. In stacks of bases oriented plane to plane, the n-π transitions are parallel to one another along the double helix, and hence this transition can be amplified by interaction. A distinctive feature of n-π transitions, their disappearance on acidification of the medium, is observed: Acid DNA has a maximum at 250 mμ in its excitation spectrum. In a weakly alkaline medium the transitions not only of guanine but also of all the other bases may interact, and in this case the band will be very sensitive to all kinds of disturbances in the secondary and tertiary structure.

The luminescence of nucleic acids in powder form shows different features. Under these conditions all the nitrogenous bases are implicated in the formation of the fluorescence spectrum of DNA (Agroskin, Korolev, et al., 1960; and Barenboim, 1963).

The ability of nucleic acids to phosphoresce is much more pronounced than their ability to fluoresce. The phosphorescence of water-glycerol solutions of nucleic acids at liquid-nitrogen temperature was observed as far back as 1957 by Steele and Szent-Györgyi. While Douzou et al. (1961) found a structureless phosphorescence band with a maximum at 475 mμ for snowlike aqueous DNA at 77° K, Bersohn and Isenberg (1964), working with water-glycerol glasses of DNA, recorded separate structural elements in the phosphorescence spectrum by using a phosphoroscope with a time resolution of 10^{-3} sec. It is characteristic that the phosphorescence spectrum in this case was formed by the addition of the phosphorescence spectra of adenine (maxima in the case of free deoxyadenosine monophosphate at 390, 415, 445, and 460 mμ) and guanine (deoxyguanosine monophosphate has a broad maximum at 420 mμ). According to Bersohn and Isenberg (1964), pyrimidines (cytosine and thymine) do not exhibit phosphorescence at 77° K but fluoresce with a maximum at 340 to 350 mμ.

In this case, however, the absence of phosphorescence is probably not equivalent to the absorption of triplet electronic excited states. The fact is that pyrimidines in the triplet state have a low π-electron density between carbon atoms 5 and 6 (Mantione and Pullman, 1964), which is equivalent to breakage of a bond. This may lead to 100% loss of excitation energy in the course of photochemical hydration or simply in the course of effective vibrations of the 5-6 bond. Theoretical considerations and some of the characteristics of DNA fluorescence led Bersohn and Isenberg to put forward the extremely attractive hypothesis of an exciton triplet-triplet mechanism of migration of the energy of the electronic excited state along the double helix of the macromolecule. One of the main experimental justifications of this view is provided by quantitative experimental investigations of fluorescence quenching by paramagnetic ions. It was found that a single paramagnetic ion can quench the phosphorescence of many bases; it "extracts" energy not only from the base to which it is directly "attached," but also from bases which are not in contact with it and are spatially distant from it. No parallel quenching of the fluorescence is observed. This indicates energy collectivization via triplet levels and not singlet levels. According to the authors' calculations, the time of localization of the exciton on an individual base is less than 10^{-9} sec. The representation of both adenine and guanine in the phosphorescence spectrum of DNA and the dependence of the half-decay time on the wavelength exclude the possibility of energy transfer between adenine and guanine, so that the exciton can appear only between molecules of the same kind. The theory of the triplet nature of delocalization of electronic excited states in DNA is greatly strengthened by the experiments of Isenberg, Leslie, et al. (1964). In an investigation of complexes of acridine dyes with DNA, these authors found three kinds of luminescence: phosphorescence of DNA, phosphorescence of the dye, and retarded fluorescence of the dye. The last kind is of greatest interest, since it is not excited by direct hits of the quanta on the dye molecule itself but is excited only indirectly by quanta absorbed by the nitrogenous bases of DNA. The intensity of the retarded luminescence with increase in the intensity of the exciting light increases linearly with the intensity of DNA phosphorescence, i.e., triplet concentration. The fact that the lifetime of the retarded luminescence was reduced from 10^{-1} to 5×10^{-2} sec with increase in the number of dye molecules per nitrogenous base led the authors to the conclusion that initially triplet-triplet migration of energy occurs between the nitrogenous bases until a "collision" with the nearest dye molecule, and then the energy migrates to it (triplet-singlet mechanism). According to the authors' estimate, the time of migration between neighboring guanine molecules along the DNA double helix is 10^{-4} to 10^{-1} sec. Rahn et al. (1964) also postulate delocalization of triplets in synthetic polynucleotides. On the other hand, delocalization of electronic excited states of nucleic acids can apparently occur via the first singlet state of purines (Weil and Calvin, 1963; and Gagloev and Vladimirov, 1964).

Chapter 5

LUMINESCENCE OF BIOPOLYMERS IN THE CELL

Soviet scientists can claim the credit for the first investigations of the natural, primary ultraviolet luminescence of cells, which is due, as has now been shown, to protein.

In 1958 Brumberg investigated the primary luminescence of biological materials with the MUF ultraviolet luminescence microscope which he had designed. The first investigations showed that cells of microbial, plant, and animal origin exhibit intense ultraviolet luminescence on excitation by the short-wave part of the ultraviolet spectrum (250 to 280 mμ). Simultaneously and independently of Brumberg, Meisel', et al., Vladimirov (1957) used another method of macroscopic investigation of the fluorescence excitation spectrum of a yeast suspension and came to the conclusion that yeast cells emit ultraviolet fluorescence which passes through a UFS-3 filter and has an excitation spectrum with a maximum at 280 mμ. He interpreted this fact as evidence that protein is the luminescence center in yeast cells. We can single out two questions around which all the investigations of primary ultraviolet luminescence of cells and tissues are concentrated.

Firstly, there is the question of the nature of the luminescence centers of the cell, i.e., the nature of the molecules whose electronic excited states are responsible for the luminescence spectrum, on one hand, and the centers which act as sensitizers of this luminescence, on the other.

Secondly, there is the question of first-ranking importance for biology: the interrelationship of the luminescence of the cell and its physiological state, the correlation between the functional or pathological state of the cell and its spectral luminescent characteristics.

We shall try to examine briefly all the experimental data now available and the conclusions derived from them for these questions separately.

NATURE OF CENTERS RESPONSIBLE FOR PRIMARY ULTRAVIOLET LUMINESCENCE OF CELLS

The question of the nature of the primary luminescence of the cell is fairly confused. In their earliest investigations Brumberg, Barskii, and Shudel' (1960) postulated that the fluorescence of cell nuclei was due to nucleic acids, and that of the cytoplasm, to RNA and proteins. However, subsequent experiments, which showed that gradual removal of free nucleotides, RNA, and DNA from the cell was accompanied by a sharp reduction of the luminescence intensity (Brumberg, Barskii, and Shudel', 1960), led to the conclusion that the luminescence was due to nucleotides. This conclusion was soon contradicted by the fact that the excitation spectra of the fluorescence of rat bone-marrow cells, liver cells, and mouse spermatids (Agroskin and Barskii, 1961) had a maximum at 280 to 290 mμ, reminiscent of that of the excitation spectra of proteins. That the luminescence was of protein nature was also indicated by the experiments of Agroskin and Pomoshchnikova (1962), who confirmed the 284-mμ maximum in the excitation spectra of yeasts and several microbes: *Saccharomyces cerevisiae* 12, *S. ludwigii* 10, *Endomyces magnusii, Escherichia coli, Bacillus anthracoides, B. mycoides, Micrococcus rubi,* and *Pseudomonas fluorescens lignifaciens.*

These experiments would probably have led to the conclusion of the exclusive role of proteins as the luminescence center of cells if it had not been known that in several cases nucleic acids and nucleoproteins have the same long-wave maxima in their luminescence excitation spectra (Konev, 1957; Agroskin, Korolev, et al., 1961; Douzou et al., 1961; and Isenberg et al., 1964).

In later investigations Brumberg and his colleagues (1961 to 1964) returned again and again to the question of the nature of the primary ultraviolet luminescence of cells. They put forward two more arguments against the protein nature of the luminescence of cells. First of all, there is no direct relationship between the tryptophan content of cells and the luminescence intensity (Barskii et al., 1962; and Pil'shchik and Nikalaeva, 1963). According to Brumberg et al. (1963), the protein nature of the luminescence is also contraindicated by the high sensitivity of this fluorescence to changes in the physiological states of the cells. This led them to "postulate the existence of a strong dependence of the latter on protein structure." The picture of cell ultraviolet luminescence became even more complicated after the publication of two papers, those of Chernogryadskaya and Shudel' (1962) and Brumberg et al. (1963A). The first of these papers showed that cell luminescence is due to an oxidized form of diphosphopyridine nucleotide. This conclusion of the authors was based on an obvious misunderstanding, since reduced diphosphopyridine nucleotide does not luminesce in the ultraviolet but in the visible region with a maximum at 465 mμ (Velick, 1961), and the oxidized form does not fluoresce at all. In the second paper, Brumberg et al. turned attention to the fact that the prosthetic group of one of the respiratory enzymes (ubiquinone or coenzyme Q) fluoresces in the reduced form in the ultraviolet region of the spectrum. A compound related to ubiquinone, hydroquinone, has a fluorescence spectrum with a maximum at 325 mμ and an excitation spectrum with a maximum at 290 mμ. The two maxima correspond fairly closely with those which are observed in cells. However, the evidence given by the authors to substantiate the implication of ubiquinone in cell luminescence is highly questionable and reduces in essence to the fact that mild treatment of cells with perchloric acid leads to a great reduction of the luminescence. On employing stronger treatment with the same reagent, the authors found that the luminescence was enhanced again. These changes in waves of luminescence were interpreted by the authors as follows: Perchloric acid, a weak oxidant, converts ubiquinone to the oxidized form, and cell luminescence is quenched, since the protein in it does not luminesce owing to strong screening by oxidized ubiquinone and nucleic acids. With further oxidation of nucleic acids and ubiquinone by perchloric acid, the screening is removed and the protein begins to fluoresce.

It is easy to see that a similar explanation would hold for the simple reason that, on the basis of the content of protein, nucleic acids, and ubiquinone in the cell and their molar extinctions, one can calculate that proteins account on the average for more than 50 to 70% of the absorbed quanta. Even in viruses, which are rich in nucleic acids, two-thirds of the absorption at 280 mμ is due to protein (Duddy, 1957). In the case of mitochondria, which contain

65 to 70% protein, 25 to 30% lipids, 0.5% RNA, and only 10^{-2}% ubiquinone, protein absorption in the 250- to 300-mμ region accounts for about 90% of the quanta. In addition, from the viewpoint of luminescence microscopy the cell can be regarded approximately as an aqueous solution containing on the average 8.5% protein and 1.5% nucleic acids, and having an optical thickness of 10 μ (0.001 cm). Taking the mean optical density of a 1% protein solution in the 250- to 300-mμ region as 5, and that of nucleic acids as 50, we obtain the cell optical density $D_{cell} = \epsilon_1 c_1 l + \epsilon_2 c_2 l = 5 \times 8.5 \times 0.001 + 50 \times 1.5 \times 0.001 = 0.120$. For such optically transparent systems there is hardly any screening of some molecules by others. In this particular case complete removal of nucleic acids would lead to enhancement of the luminescence by only 7.5%. Hence, there cannot be any significant screening effect of nucleic acids or ubiquinone, leading to extinction of the cell luminescence.

In their recent paper Shudel' et al. (1964), who certainly obtained interesting data on the effect of respiratory inhibitors (potassium cyanide, sodium amytal, and dinitrophenol) on Ehrlich's ascites carcinoma cells, nevertheless complicated the question of the fundamental nature of the cell emitters even more by including tocopherol and vitamin B_{12} among them. Whatever may be the case, there are now several different points of view on the nature of the substances responsible for the ultraviolet luminescence of cells and its elements. These substances include the luminescent form of DNA and proteins in nuclei and proteins, RNA, free nucleotides, including diphosphopyridine neucleotide, reduced ubiquinone, tocopherol, and vitamin B_{12} in the mitochondria. A more accurate qualitative and quantitative picture of the mechanisms of cell ultraviolet luminescence can obviously be obtained in two ways: by extracting luminescent molecules from the cell and identifying them and by investigating the greatest possible number of spectral characteristics of the primary luminescence of the undamaged cell which might help to establish the chemical nature of the luminescence carriers.

The first question which can reduce the uncertainty is: Are one or several different molecular centers responsible for the ultraviolet luminescence of the cell?

An answer to this very important question can be directly obtained by investigating the fluorescence spectra of the native intact cell. That the fluorescence spectra of different cells and their main elements (mitochondria and nuclei) consist of a single

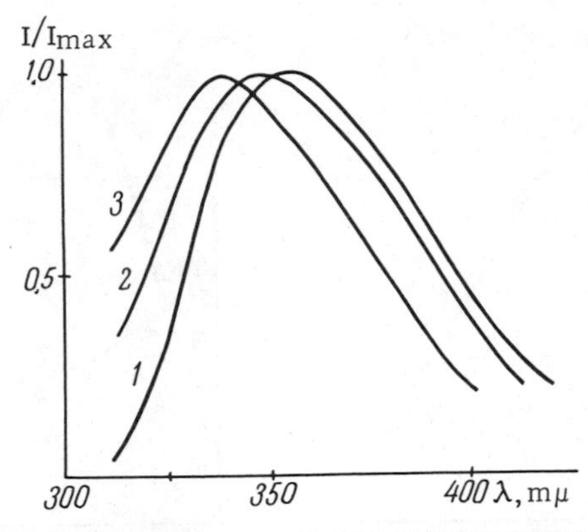

Fig. 51. Fluorescence spectra at room temperature: (1) aqueous solution of tryptophan (concentration 10^{-4} mole/liter), (2) mitochondria in physiological saline, (3) nuclei in physiological saline (Konev, Lyskova, and Bobrovich, 1963).

structureless band suggests at first glance that there is only one kind of fluorescent molecule in the cell (Fig. 51). Actually this is not the case at all. A single structureless fluorescence band can in principle be the composite of a set of closely situated fluorescence spectra of several different molecular centers.

A conclusive answer requires comparison of the family of fluorescence spectra of cells excited with monochromatic light of different wavelengths. In the case of a single-center luminescence mechanism, all the spectra will be the same, whereas, in the case of a many-centered mechanism, the contribution of different molecules to the total fluorescence spectrum will change from one exciting wavelength to another, and, as a consequence, the fluorescence spectrum of the cells will be altered (if there is no energy migration with 100% efficiency).

It has been found that the fluorescence spectra of several objects of microbial and animal origin (yeast cells and *E. coli* cells, cells of the esophagus, muscles, lungs, and brain of the frog, mitochondria and nuclei of white-rat liver) depend only slightly on the wavelength of the exciting light in the range of 230 to 300 mμ (Fig. 52). These experiments show that only one kind of molecule is implicated in the formation of the fluorescence spectrum. The same conclusion can be derived from the fact that the phosphorescence spectrum of biological objects is independent of the exciting wavelength in the range of 250 to 296 mμ. It is

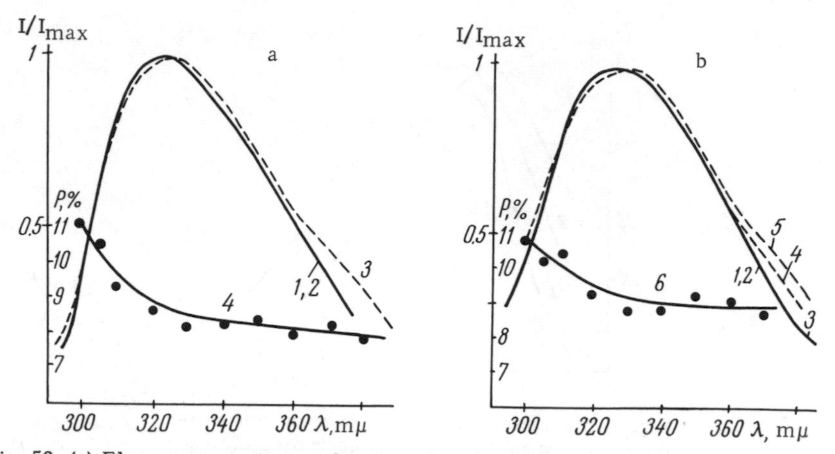

Fig. 52. (a) Fluorescence spectra of nuclear fractions in sucrose at room temperature on excitation by (1) 265, (2) 280, and (3) 302 mμ and (4) emission polarization spectrum of fluorescence of nuclei excited by 265 mμ; (b) fluorescence spectra of mitochondrial fractions in sucrose at room temperature on excitation by (1) 265, (2) 280, (3) 296, (4) 254, and (5) 302 mμ and (6) emission polarization spectrum of fluorescence of mitochondria excited by 265 mμ (Konev, Lyskova, and Bobrovich, 1963).

characteristic that even nuclei, which are rich in DNA, have only one kind of luminescent molecule available for excitation: Their fluorescence spectra and even their phosphorescence spectra at liquid-nitrogen temperature are independent of the wavelength of the exciting light.

What is the nature of this single luminescence center? A reliable answer to this question can be obtained by investigating the various spectral luminescent characteristics of the cell.

Figure 53 shows the fluorescence excitation spectra of nuclei and mitochondria of white-rat liver. These spectra are practically identical with the excitation spectra of proteins. The excitation spectra distinctly show the long-wave (280 mμ) and short-wave (227 to 230 mμ) absorption maxima of aromatic amino acids in protein. At wavelengths greater than 300 mμ, excitation of luminescence is weak. Yeast and *E. coli* cells exhibit similar excitation spectra.

The appearance of the two absorption bands of protein-bound tryptophan in the fluorescence excitation spectra indicates definitely the protein nature of the primary luminescence of cells. The values of the fluorescence quantum yields also indirectly indicate the protein nature of the luminescence centers (Konev, Lyskova, and Bobrovich, 1963). The high quantum yields (5 to 25%) of the fluorescence of rat-liver mitochondria and the cells of most frog tissues indicate that the luminescence center cannot be a substance present in the cell in microquantity, such as coenzymes or vitamins, since their share of the total number of quanta absorbed by the cell would only be hundredths or even thousandths. The high yields show that the main light-absorbing substance in the cell, i.e., protein, is responsible for the luminescence (Table 17). The τ of fluorescence (3.5×10^{-9} sec) and τ of phosphorescence (5.9 to 6.0×10^{-9} sec) of mitochondria and nuclei of white-rat liver have values typical of tryptophan-containing proteins. The fluorescence polarization spectra (Fig. 54) also indicate the protein-bound tryptophan nature of the primary luminescence of cells. The absorption polarization spectra of fluorescence, like that for tryptophan-containing proteins, show a maximum at 265 to 270 mμ, a dip in the 280- to 285-mμ region, and a considerable increase in polarization in the long-wave region of 296 to 300 mμ. The absolute values of the polarization (9 to 14% for the maxima at 270 mμ) also agree with those for protein. This rules out the case in which energy migration occurs with the luminescence due to some nonprotein center which collects energy from protein molecules as a result of transfer. In fact, the fluorescence after energy transfer would be completely depolarized, and a protein form of polarization spectrum would be impossible in such a case. At the

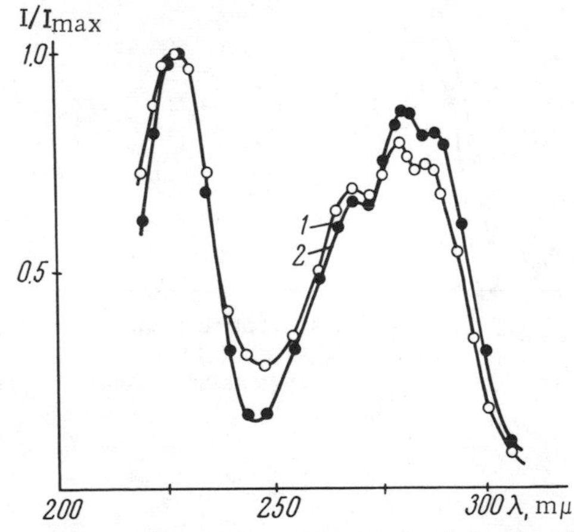

Fig. 53. Excitation spectra of fluorescence of mitochondria (1) and nuclei (2) in physiological saline at room temperature (Konev, Lyskova, and Bobrovich, 1963).

Table 17. Quantum Yields of Fluorescence of Various Frog Tissues

Material	Quantum yield, %, λ_{exc} = 265 mμ			Quantum yield, %, λ_{exc} = 296 mμ
	Just after preparation	After 30-min treatment with 8 M urea	After freezing 10 times	Just after preparation
Muscle (gastrocnemius)	15.0	17.0	15.0	25
Tongue	7.3	7.3	6.9	23
Esophagus (epithelium)	8.2	5.9	8.0	16
Liver	5.2	4.2	5.8	18
Lungs	5.9	4.2	5.1	17
Brain (hemisphere)	9.0	...	...	15
Heart	5.4			

same time, the polarization spectra clearly demonstrate that the two oscillators (^1L_a and ^1L_b) of the long-wave absorption band of tryptophan are involved in the absorption and emission of light.

The emission polarization spectra of the fluorescence of mitochondria and nuclei have a distinctly protein shape (Fig. 52). The fluorescence spectra are characterized by a constant and fairly

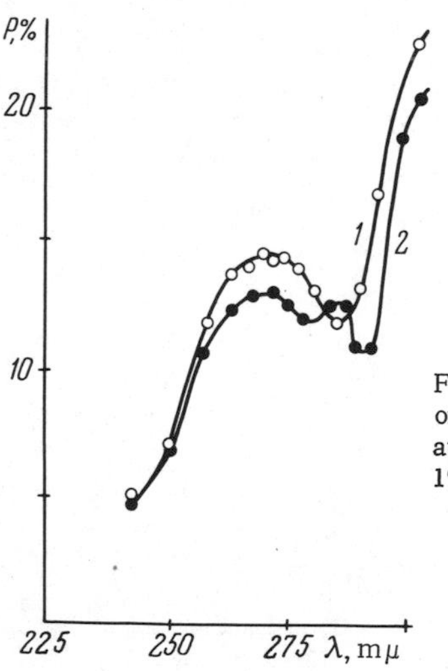

Fig. 54. Absorption polarization spectra of fluorescence of mitochondria (1) and nuclei (2) in physiological saline at room temperature (Konev, Lyskova, and Bobrovich, 1963).

low polarization (about 10%) for the whole long-wave profile of the band and an increase of the polarization to 13 to 14% in the short-wave region. This type of emission polarization spectrum of fluorescence is typical of tryptophan-rich proteins in which inter-tryptophan energy migration occurs with the participation of the ^1L_a electronic level. The constancy of the fluorescence polarization in the range of 310 to 400 mμ and the fact that the fluorescence spectra are independent of the exciting wavelength indicate the presence of only one luminescence center: The luminescence oscillator at all points of the spectrum is oriented in space at the same angle to the absorption oscillator. The tryptophan nature of the luminescence is very clearly expressed in the low-temperature luminescence spectra of mitochondria. A cursory glance at Figs. 55 and 56 is sufficient to see that the typical tryptophan "trident," which is difficult to confuse with anything else, is clearly revealed in the low-temperature luminescence spectra. This trident, together with the shoulder at 480 mμ, is characteristic of native nuclei and mitochondria of various tissues. The polarization of the phosphorescence according to the emission spectrum is constant and has small negative values (-5%). All this excludes the implication of centers other than tryptophan in the low-temperature luminescence.

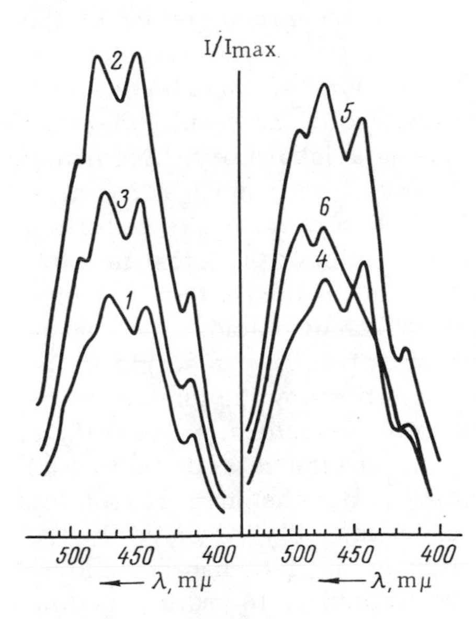

Fig. 55. Phosphorescence spectra of nuclei in physiological saline at liquid-nitrogen temperature in relation to wavelength of exciting light; exciting wavelength in millimicrons: (1) 265, (2) 270, (3) 280, (4) 292, (5) 296, and (6) 302 (Konev, Lyskova, and Bobrovich, 1963).

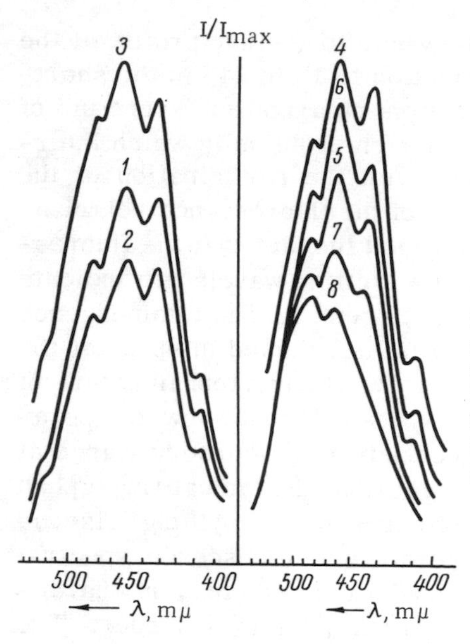

Fig. 56. Phosphorescence spectra of mitochondria in physiological saline at liquid-nitrogen temperature in relation to wavelength of exciting light; exciting wavelength in millimicrons: (1) 265, (2) 270, (3) 280, (4) 289, (5) 292, (6) 296, (7) 302, and (8) 313 (Konev, Lyskova, and Bobrovich, 1963).

Removal first of free nucleotides and RNA from the cells (treatment of mitochondria and nuclei with 10% perchloric acid for 18 hr at room temperature) and then of DNA (treatment with hot 5% perchloric acid for 40 min) does not lead to any appreciable changes in the above-mentioned spectral characteristics of the luminescence. This again confirms the largely protein (tryptophan) nature of cell luminescence. The results of experiments with hydrochloric acid in our laboratory contradict the results obtained by microscopic techniques in Brumberg's laboratory. According to the earlier results of this laboratory, the gradual removal of free nucleotides, RNA, and DNA leads to a progressive reduction of the fluorescence. According to the results of the latest investigations, the fluorescence is quenched at first and then becomes enhanced by stronger treatment with perchloric acid. The reason for disagreement of the results should probably be sought in the fact that in the case of luminescence microscopy the luminescence intensity is affected by several subsidiary factors, primarily the partial removal of tryptophan and the change in cell volume, as Agroskin and Barskii (1961) indicated. But the main reason for the "play" of the cell luminescence intensity detected by the luminescence-microscope method lies in the fact that perchloric acid greatly increases the ability of tryptophan to undergo photochemical oxidation. In macroscopic experiments with cell suspensions the fluorescence-exciting light is greatly reduced on passage through the monochromator. In the ultraviolet luminescence microscope a very powerful light flux is focused on the cell, and this promotes the photodestruction of tryptophan in the cell. Acceleration of the photooxidation of tryptophan in the free state and in proteins or nucleic acid by perchloric acid has been detected in preparations irradiated at the focus of an SVD-120A lamp.

For instance, the quantum yield of the fluorescence of nuclei falls from 2 to 1% after treatment with hot perchloric acid and after a 1-min exposure at the focus of an SVD-120A lamp is reduced to one-sixth of the latter value (Table 18).

Table 18. Quantum Yields of Fluorescence of Cell Elements (Konev, Lyskova, and Bobrovich, 1963)

Object and conditions	Quantum yield	
	280 mμ	265 mμ
Tryptophan in distilled water	0.22	0.22
Tryptophan in distilled water after irradiation at focus of SVD-120A lamp for 1, 2, 3, and 4 min*	0.22; 0.16; 0.16; 0.16	0.22; 0.16; 0.16; 0.16
Tryptophan after treatment with $HClO_4$ and irradiation at focus of SVD-120A lamp for 1, 2, 3, and 4 min	0.18; 0.12; 0.09	0.18; 0.12; 0.09; 0.09
Tryptophan mixed with mitochondria	0.30	0.30
Native nuclei in physiological saline	0.02–0.03	0.01–0.02
Treated nuclei	0.01–0.015	0.008–0.015
DNA-free nuclei after 1-min irradiation with SVD-120A lamp	0.0015–0.0025	
Mitochondria in physiological saline	0.10–0.15	
Mitochondria in 8 M urea	0.11–0.17	
Mitochondria after boiling	0.11–0.17	

*The solution was irradiated in a total-absorption layer in a closed cuvette with restricted access of atmospheric oxygen.

The photochemical breakdown of tryptophan and proteins proceeds much more slowly in the absence of perchloric acid.

Thus, from the point of view of the protein nature of the luminescence of cells and its elements, only the two closely connected facts observed by Brumberg and mentioned at the beginning of this section require explanation.

Firstly, why is there not a direct relationship between the luminescence intensity and tryptophan content of the cell,. and, secondly, why are there wide variations in the intensity of this luminescence? The fluorescence intensity of the cell is not proportional to the amount of tryptophan contained in it for several different reasons: changes in the relative amount of active and inactive absorption, i.e., differences in the degree of screening of tryptophan by other molecules (e.g., nucleic acids, hemoglobin etc.), the light-scattering properties of the cell medium, the actual assortment of individual proteins in the particular cell, each of which has its own very variable quantum yield, and, finally, a very important point, the nature of the supermolecular packing of the protein, which, as was shown earlier in the case of milk casein, can alter the quantum yield of the fluorescence by a factor of 2 or more. The combined and sometimes conflicting action of all these four factors must destroy the proportional relationship between the fluorescence intensity and the tryptophan content of the cell, although, as a whole, there should definitely be some correlation. This is indicated by the data of Brumberg et al. (1958), who observed a marked enhancement of the luminescence of yeast cells fed with tryptophan. All the points mentioned above enable the tryptophan luminescence of cells to respond to the functional-physiological or -pathological changes in the cells and in the organism.

Thus, the whole body of experimental information indicates the protein nature of the ultraviolet luminescence of cells. At the same time, the hypothesis of the nucleotide nature of the luminescence, even with a correction for the existence of two physicochemical forms of nucleic acids, luminescent and nonluminescent (Brumberg and Barskii, 1960), is not confirmed by experiment. The only example which these authors give in support of the existence of a luminescent form of nucleic acid is that of the large chromosomes in the salivary-gland cells of the *Chironomus* larva. This object showed a strong fluorescence of the DNA-rich chromosome disks and a weak fluorescence of the interdisk regions, which contain mainly protein. This example, however, is a rather dubious one. First of all, this observation does not in any case justify the assumption of intravital occurrence of a luminescent form of DNA, since the luminescence was detected in salivary-gland cells fixed in Carnoy's fluid. Under these conditions there could have been artifacts due to luminescent products of the chemically altered chromosome material. In addition, the chromosome disks caused greater blackening of photographic plates than the interdisk regions in transmitted light (250 to 280 mμ), i.e., they exhibited greater absorption. In the absence of any quantitative characteristics of the light absorption and the luminescence intensity, the relatively greater intensity of luminescence of the chromosome disks could have been due simply to their greater absolute protein content, which manifests itself, despite screening, by nucleic acids. In any case, the question of the luminescence of this object is not clear and merits more careful investigation.

All animals and microbial cells have the same common universal mechanism of ultraviolet luminescence; in every case the luminescence is due to singlet electronic excited states of the tryptophan of proteins. The situation is a little different in the case of plant cells, which are rich in cellulose. It has been shown (Konev, Bobrovich, and Lyskova, 1965) that in cells of the root system of cereals the protein maximum of the fluorescence spectra at 340 mμ appears in pure form only on monochromatic excitation by light of wavelength 280 and 220 mμ, i.e., on selective excitation at the absorption maxima of protein aromatic amino acids. For other exciting wavelengths the protein luminescence is masked to some extent by a broad diffuse band with a maximum at 430 mμ. This fluorescence has a broad excitation region, from 230 to 365 mμ, and is very probably due to cellulose. Hence, in the case of plant cells, which are rich in cellulose, we are dealing with two centers responsible for the ultraviolet luminescence and not with one, as in the case of animal and most microbial cells.

The next stage in the investigation of cell ultraviolet luminescence after the establishment of its protein nature was to study the physicochemical energetic state of protein tryptophan in its natural intravital environment. In view of the high sensitivity of tryptophan to the properties of the medium, to the microenvironment, we could infer a priori that tryptophan may act as a kind of indicator lamp reflecting perturbations in the molecular organization of the cell. Thus, how is the luminescence affected by the incorporation of tryptophan residues along with a protein carrier in biological structures? First of all, do the proteins of the cell and its structures manifest the two singlet electronic excited states of protein tryptophan, one of the most characteristic features of the energetics of this aromatic amino acid?

As mentioned above, the most reliable evidence of the existence of one or two singlet electronic excited states of proteins in the cell can be obtained from an investigation of the emission polarization spectra of the fluorescence of the cell. Figure 52 shows that the fluorescence of mitochondria and nuclei has constant polarization values in the middle- and long-wave part of the band. On approach to the short-wave end of the fluorescence band there is a quite definite increase in polarization: Nuclei and mitochondria have a polarization of 9 and 8%, respectively, in the main part of the fluorescence spectrum, but in the 300-mμ region the polarization increases to 11 to 12%. Similar emission polarization spectra have been obtained for a suspension of live yeast cells. Since the nature of the emission polarization spectra of fluorescence is not appreciably altered after fixation of the cell, the relative probabilities of radiation from the ^1L_a and ^1L_b electronic levels do not depend on the structural organization of the cell or the nature of metabolism in it. In general, as in the case of incorporation of tryptophan residues in proteins, further incorporation of the protein itself in supermolecular biological systems has little effect on the polarization spectra. Hence, the number and relative orientation of the main absorption, fluorescence, and phosphorescence oscillators in the cell are the same as in tryptophan residues in the free state. Slight variations are observed only in the relative intensities of the vibrational sublevel at $\lambda = 285$ mμ of the ^1L_b electronic level. It is characteristic that for the total pool of protein molecules in the nuclei this sublevel has a slightly greater intensity than for the total pool of protein molecules in mitochondria. This is manifested in the appearance of a second minimum in the polarization spectra. This fact indicates that the physicochemical energetic state of tryptophan in the composition of different biological systems may be different. The difference in energy state of the tryptophan of nuclei and mitochondria is distinctly manifested in the position of the fluorescence band. In freshly extracted mitochondrial preparations the position of the fluorescence maximum varies in the range of 335 to 350 mμ in different animals. In contrast to mitochondria, nuclei have a constant and stable maximum at 333 mμ. In other words, the fluorescence spectrum of nuclear tryptophan is shifted in the short-wave direction in comparison with the fluorescence spectrum of mitochondria. This indicates that tryptophan residues in nuclei are surrounded by a less polar medium than in mitochondria.

The positions of the fluorescence maxima for different frog tissues also differ appreciably. Despite the fact that in all tissues the luminescent molecules are the same from the chemical point of view (tryptophan residues), the data of Table 19 show that the fluorescence spectra exhibit some tissue (organ) specificity. A particular kind of tissue or organ is characterized by its own fluorescence maximum, the position of which varies only slightly from individual to individual. Esophageal epithelium gives the shortest-wave fluorescence maximum, and tongue cells give the longest-wave maximum.

Differences between tissues are also revealed by a comparison of the relative intensity of the singlet and triplet luminescence.

Table 19. Position of Luminescence Maxima and Their Relative Intensity in Frog Tissues and Some Microorganisms (Konev, Bobrovich, and Lyskova, 1965)

Material	$\lambda_{max.\ fl}$, mμ				Ratio S/T
	Native state, t=20°C	8 M urea, t=20°C	Water extract, t=20°C	Native state, t=−196°C	
Tongue	330, 330, 330*	340	340	320	2.35; 1.9
Lungs	323, 323, 323	340	340	319	3.1; 3.2; 3.36
Heart	321, 320, 322	340	342	319	3.2; 3.0; 3.1
Brain	324, 324, 324	340	340	319	2.6; 2.57
Esophagus	320, 320, 320	340	340	319	3.0; 3.5; 2.5
Liver	328, 328, 328	340	350	320	2.35; 3.25; 2.30; 2.20
Gastrocnemius muscle	325, 325, 325	340	340	321	1.8; 2.1; 1.8
Yeast suspension	324, 324			320	1.55; 1.40
E. coli suspension	340			331	1.75; 1.00

*Values for three different individuals are given.

Table 19 shows that the ratio of the fluorescence intensity to the phosphorescence intensity at their maxima (S/T) differs considerably in cells of different tissues.

Hence, the "average" tryptophan residue in cells of different tissues is situated in a microenvironment which differs in its polarity and degree of hydrogen bonding of the amino group. There may be two reasons for this: differences in the primary structure of the protein macromolecules, i.e., differences in protein composition, or differences in the secondary, tertiary, or quaternary structures.

The structural nature of the cause of the difference in the position of the fluorescence maxima is particularly well illustrated by experiments involving treatment of cells of different tissues with 8 M urea. It is known that urea does not cause any changes in the fluorescence spectrum of tryptophan itself and yet is capable of altering the position of the fluorescence maxima of proteins by destroying the secondary and tertiary structures by denaturation. The position of the fluorescence maximum of different proteins is stabilized at 350 mμ. This means that 8 M urea can be used as a kind of "developer" to reveal the factor causing intertissue variation of the position of the fluorescence bands. As the table shows, incubation of different tissues for 30 min in 8 M urea causes different long-wave shifts of the fluorescence spectra. As a result, the position of the fluorescence maxima of different tissues becomes 340 mμ for heart, brain, lungs, esophagus, skeletal muscles, and also liver.

Hence, differences in the protein structures are the direct cause of intertissue differences in fluorescence spectra.

This experiment, however, does not enable us to discriminate between the above two possibilities, the differences in the average set of proteins or the differences in the supermolecular structurization of the same proteins, since urea can not only denature proteins but also destroy supermolecular structures produced by hydrogen bonds or other weak forces. A choice between these possibilities can be made from a comparison of the fluorescence spectra of the pool of protein molecules of cells of different tissues in the free state, in solution. If the proteins of different tissues in the free state still show spectral differences, it is a matter of the first cause, different protein composition of different cells. If free proteins of cells of different tissues give the same fluorescence spectra, it will indicate that the previously observed differences are a consequence of the different kind of packing of the proteins in supermolecular associations.

An investigation of the fluorescence spectra of saline extracts of various organs has shown that in this case differences in the position of the fluorescence bands practically disappear. For these tissues there is a maximum at 340 mμ, except for liver, where the maximum is shifted to 350 mμ. Hence, the total pool of protein molecules in cells of different tissues and organs does not differ in composition (from the viewpoint of the luminescence method), except for liver, but has different supermolecular structure. We can surmise that hydrophobic interactions between neighboring protein macromolecules transfer superficially situated tryptophan residues from a hydrophilic to a hydrophobic microenvironment, and this in turn leads to a short-wave shift of the fluorescence spectra.

LUMINESCENCE AND FUNCTIONAL STATE OF THE CELL

An investigation of the primary ultraviolet luminescence of cells would have very little practical or theoretical significance for biology if three groups of facts had not been established: (1) Cells of different tissues differ in their luminescence; (2) the luminescence is altered by a change in the functional physiological state of the cell; and (3) conversion of the cell to the pathological state is accompanied by corresponding changes in luminescence.

We shall discuss briefly the facts which confirm these points.

1. It was reported in a paper by Barskii et al. (1959) that the meristematic tissue in the hyacinth ovary luminesces more strongly in the 340- to 380-mμ region than the cotyledon cells. Intense luminescence of meristematic tissues has also been observed in other botanical material.

Great differences in the luminescence intensity are characteristic of white blood cells (Brumberg, Barskii, Kondrat'eva, Chernogryadskaya, and Shudel' 1961). The cytoplasm of myeloblasts, myelocytes, and metamyelocytes fluoresces most intensely. Lumphocytes of peripheral blood and bone marrow fluoresce weakly.

According to Khrushchev (1964), histiocytes, fat cells, leucocytes, and undifferentiated (cambial) cells of connective tissue luminesce strongly. Fibrous structures of loose, unformed connective tissue and cells of the fibroblastic series fluoresce weakly.

Differences in fluorescence intensity of different frog tissues have been observed in investigations in our laboratory. Cross-striated muscle is characterized by the greatest luminescence intensity. If its intensity is taken as unity, the luminescence intensity of the cerebral hemispheres is 0.6, esophageal epithelial cells 0.53, tongue 0.55, lungs 0.4, liver 0.30, and heart 0.35. Of much greater interest than the establishment of differences in the luminescence intensity of cells of different tissues is the investigation of the luminescence spectra. Accepting the exclusively protein nature of the primary luminescence of the cell, we can link the changes in the luminescence intensity of cells of different tissues with variations in the ratio of active (protein) and inactive (other substances, mainly nucleic acids) absorption. Differences in cell fluorescence spectra, which are due to the same center, tryptophan, can be attributed entirely to the fact that the tryptophan residues are in different physicochemical states.

2. The first measurements of fluorescence spectra (Brumberg, Meisel', et al., 1958, using the photographic method with ultraviolet luminescence microscope) showed that the fluorescence spectra of tissues depended on their age. In an adult mouse, collagen fibers in a cross section of the tail gave an intense short-wave luminescence with a maximum at about 313 mμ, probably due to tyrosine, which has a maximum at 304 mμ according to a more accurate determination. In a young mouse, however, the collagen fibers did not give such fluorescence. The authors correctly attributed these differences in the spectra to changes in the chemical composition of the tissues with age. Other observations with the luminescence microscope showed quenching of the fluorescence of rat bone-marrow cells when they died in physiological saline (Brumberg, Barskii, Kondrat'eva, Chernogryadskaya, and Shudel', 1961). A definite conclusion regarding this observation is very difficult, since the trivial mechanism of release of the cell substance (protein) into the physiological saline on cell death is possible. Of much more interest are the observations of Pil'shchik and Nikolaeva (1963) regarding the age-dependent changes in the fluorescence intensity of white-rat liver cells. In a whole series of replicates they managed to find a regular variation of the fluorescence intensity of cells during embryogenesis. The weak fluorescence intensity of cells on the fourteenth

to fifteenth days of development is followed by an enhancement on the sixteenth day. The first wave of fluorescence is then replaced by a reduction on the seventeenth to eighteenth days and a second wave of enhanced fluorescence on the nineteenth to twentieth days. On the twenty-first day of embryogenesis and on the first day of postnatal development, the fluorescence intensity is greatly reduced again. These data indicate regular qualitative changes in the cells at the molecular level during embryonic development. Bresler et al. (1964) arrived at a more definite interpretation of the nature of the periodic changes in the luminescence intensity of liver cells. In their experiments they investigated the time course of the changes of several luminescence and cytological characteristics of rat-liver cells in the process of regeneration after the removal of approximately 65% of the liver tissue. The ultraviolet luminescence intensity during regeneration varied in a manner similar to that of embryogenesis. The authors came to the conclusion that the "change in intensity of the ultraviolet fluorescence of regenerating liver cells is due to a change in the amount and structure of its proteins."

Among this group of observations on age physiology we can include the data of Rozanova et al. (1963) on the more intense fluorescence of lymphoblasts from lymph nodes than of mature lymphocytes of peripheral blood and of erythroblasts (low intensity) in comparison with erythrocytes, which do not fluoresce at all.

The last fact has a very obvious explanation if we consider that erythrocytes consist mainly of hemoglobin, in which the fluorescence of the globin tryptophan is quenched as a result of migration to heme of the energy absorbed (Teale, 1959).

Several conclusions regarding the relationship between the ultraviolet luminescence of cells and their differentiation during ontogenesis can be derived from the luminescence-microscope observations of Khrushchev (1964). He observed that the fluorescence of fibroblasts in mouse connective tissue decreases in the process of differentiation. In mature fibroblasts the perinuclear regions of the cytoplasm give the highest luminescence intensity.

There is now a vast body of information indicating far-reaching chemical and physicochemical changes in the cell during preparation for mitosis and during the different stages of mitosis itself. A good confirmation of this is the results of investigation of the ultraviolet luminescence intensity of cells during mitosis (Brumberg, Meisel', et al., 1961). These authors, using a subcultured line of human amnion cells, a primary culture of guinea pig and monkey cells, and a culture of embryonal epithelium and human fibroblasts, showed that the weak luminescence of amnion cells and fibroblasts in interkinesis is increased in early prophase and reaches a maximum by the middle stage of mitosis, metaphase. The luminescence then slowly declines. From a qualitative assessment the changes in luminescence intensity correspond to changes in light absorption at 250-270 mμ. This suggests that the changes in fluorescence intensity indicate a burst of biosynthesis of a tryptophan-containing protein. This is all the more likely since it has been shown by a macroscopic technique that in a suspension of synchronized yeast cells under total-absorption conditions no changes can be found in the fluorescence intensity of cells preparing for and entering the budding state (Konev, Nisenbaum, and Lyskova). It is probably correct to assume that not only such decisive events in the life of the cell as division are expressed in luminescence. We can surmise that luminescence responds to quantitative and qualitative changes in metabolism.

In fact, a few observations in this area have recently appeared. Brumberg, Barskii, Chernogryadskaya, and Shudel' (1963) report that the intensity of the ultraviolet fluorescence of adult-rat liver cells depends on their diet. It is not clear, however, if these changes are connected with the quantity or nature of the proteins.

In the experiments of Konev and Prokopova the quantum yield of the fluorescence of a thick smear of rat-liver mitochondria during fasting was investigated. We observed a reduction of quantum yield by 10 to 15% for light of wavelengths 265 and 280 mμ in total-absorption conditions on the fifth to seventh days of fasting. This parallelism in the reduction of the quantum yield shows that the effect is not due merely to an increase in the screening effect of nucleic acids but is more probably due to qualitative changes in the pool of protein molecules.

Khrushchev (1964) observed differences in the luminescence intensity of the same white blood cells, lymphocytes, according to whether they were present in connective tissue or peripheral blood. In this case there was a distinct correlation between the luminescence of the cell and its metabolism. Regular changes in fluorescence intensity have also been observed in leucocytes during the phagocytosis of microbes (Brumberg and Brumberg, 1964).

Segment-nuclear leucocytes of mouse white blood cells which had ingested microbes had 20% greater luminescence intensity on the average than nonphagocytizing cells. The main mass of the cytoplasm of phagocytizing leucocytes luminesced approximately with its former intensity, but the regions of the cytoplasm lying close to the microbe undergoing digestion luminesced very brightly. Probably the most obvious and also the most interesting explanation of this enhancement of luminescence is that it indicates a wave of conformational changes in working protein molecules due to enhanced enzymatic activity in these regions of the cytoplasm—proteolysis, oxidative processes, etc.

The experiments of Konev and Lyskova (1965) showed that the addition of succinic or citric acid to a suspension of fasting mitochondria, which had been kept for 2 or 3 hr in physiological solution devoid of respiration substrates, leads to a gradual reduction in the fluorescence quantum yield of mitochondria (Fig. 57). The fluorescence spectrum is unaltered and is of the typical protein type with a maximum at 340 mμ.

The effect is not observed in the case of mitochondria which are vitally inactive owing to prolonged fasting or heat. In these cases the quantum yield remains constant when succinic acid is added, which rules out direct interaction between proteins and succinic acid as the cause of the change in quantum yield. This indicates that succinic acid acts by a biological pathway, via biological oxidation processes.

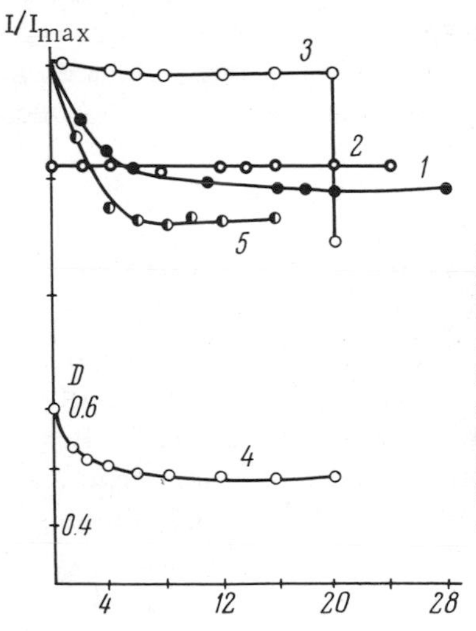

Fig. 57. Changes in relative quantum yield of fluorescence (excited by 265 mμ) on addition of drops of 1% succinic acid to 3 ml of mitochondrial suspension at room temperature: (1) freshly isolated ("living"); (2) heat-fixed; (3) rendered vitally inactive by prolonged fasting (24 hr in physiological saline); (4) changes in light scattering for curve 1; (5) the same as 1, but on addition of citric acid instead of succinic acid (Konev and Lyskova, 1965).

In this connection we must discuss Ungar and Romano's investigation (1962), which led them to formulate an extremely important conclusion for molecular biology: The structure of protein changes during the occurrence of one of the main physiological processes, conduction of excitation along a nerve.

The conduction of nerve impulses along a nerve at a repetition rate of 30, 60, 100, and 180 sec^{-1} is accompanied by a reversible reduction of 30% in the fluorescence intensity. The shapes of the excitation spectra and of the luminescence spectra, and also the fact that the luminescence centers did not pass out of the nerve extract into the physiological saline used for dialysis, indicate the purely protein nature of the luminescence. It is remarkable that a similar reduction in the fluorescence intensity of nerve proteins was produced by 6 M urea. This led the authors to advance the highly attractive hypothesis that nerve proteins participate directly in the conduction of excitation by undergoing reversible alterations of secondary and tertiary structure similar to denaturation changes.

Alteration of protein macromolecules in the excited nerve has also been shown by several authors by methods independent of luminescence: from the increase in number of free sulfhydryl groups capable of ionization (results of amperometric titration, Ungar and Romano, 1958), and from the changes in the ultraviolet absorption spectra of proteins (Ungar, Aschheim, et al., 1957).

Shtrankfel'd (1964A) attempted to detect changes in the protein fluorescence of muscle during its functioning. She observed a fourfold reduction of fluorescence intensity in frog sartorius undergoing tetanic contraction. During thermal contracture (according to her data) the fluorescence intensity was reduced at first and subsequently increased; on treatment with 2 to 10% urea it was doubled.

CELL LUMINESCENCE AND THE PATHOLOGICAL PROCESS

Published information suggests that in a number of cases luminescence is a fairly sensitive indicator of pathological states of the cell. The luminescence intensity of cells is appreciably altered when they are damaged by ultraviolet rays or ionizing radiation. Brumberg and Barskii (1960) observed a reduction of the ultraviolet fluorescence and the appearance of blue fluorescence of cells in different animal tissues when they were irradiated with intense fluxes of ultraviolet light or X rays. The long-wave luminescence was presumably due to products of the photochemical oxidation of proteins or other cell components. This was indicated, for instance, by the reduction of the effect in the presence of such a strong oxygen acceptor as hyposulfide.

According to the results obtained in our laboratory, irradiation of tryptophan in polyvinyl alcohol leads to the appearance of photochemical products with fluorescence maxima at 420 and 440 mμ. Brumberg et al. (1960), Khan-Magometova et al. (1960), Barenboim et al. (1961), and Pinto (1962) found that cell luminescence was very sensitive even to relatively small doses of X rays, where other methods could hardly detect any changes in the physicochemical properties of protein or nucleic acids. The experiments of Khan-Magometova et al. (1960) showed that the luminescence intensity of different tissues and organs of the white rat increases within 3 to 4 hr after exposure to 1000 r of X rays. Taking into consideration the fact that, as was shown earlier, the luminescence of animal tissues is due exclusively to tryptophan-containing proteins, we can postulate that in this case the observed effects reflect the changes in the fluorescence quantum yield of tryptophan of proteins which have undergone some structural changes. The constancy of the absorption coefficient of the blood plasma of irradiated animals with the simultaneous increase in fluorescence intensity also fits in well with the idea of the structural nature of the observed effects. Hence, there are some grounds for linking the observed individual response of tissues to the same treatment with differences in the level of structurization. Bone marrow showed the greatest increase in fluorescence intensity, by a factor of 1.5 after 4 hr and 1.8 after 24 hr, whereas for spleen the factor of increase was 1.2 and 1.5 and for liver it was 1.0 and 1.05.

Barenboim et al. (1961) found physicochemical changes in the cell 0.5 to 1 hr after irradiation with very small doses (only 42 rad). The relative reduction in fluorescence intensity of blood leucocytes after these doses was replaced by an increase when the dose was raised further. The maximum enhancement was observed at 336 to 756 rad.

Thus, there are grounds for believing that the changes in ultraviolet luminescence intensity associated with the radiobiological effect of small doses reflect the changes occurring at the level of supramolecular protein structures. The conformational nature of the changes in luminescence is also indicated by the broadening of the fluorescence spectra observed by Khan-Magometova et al. and by Barenboim et al. A similar broadening of the fluorescence spectra is produced by conformational changes in protein due to denaturation by heat or urea in vitro and in vivo (Konev, Lyskova, and Bobrovich, 1963).

Of great interest are the experimental observations of the luminescence of cells and tissues in such pathological states as carcinogenesis, inflammation, psychic diseases, and muscle atrophy due to denervation. The first information regarding the ultraviolet luminescence in the case of pathological conditions in the organism was obtained in 1958 (Konev). Braines et al. (1959) observed changes in the shape of the fluorescence excitation spectrum of blood serum in a group of chronic alcoholics. Changes in luminescence similar to those in schizophrenia occur in the state of experimental catatonia in dogs.

Shtrankfel'd (1964) discovered a two-phase change (enhancement followed by reduction) in the fluorescence intensity of frog sartorius after its denervation. On the hundredth day after the operation the fluorescence intensity of the denervated muscle was 30 to 35% lower than in the control, the nondenervated muscle of the other leg.

Brumberg, Barskii, Kondrat'eva, Chernogryadskaya, and Shudel' (1960) observed that in acute leucosis there was a sharp increase in the luminescence of peripheral blood lymphocytes, whereas other kinds of malignant growth were accompanied by enhancement of the fluorescence of segment-nuclear leucocytes. Rozanova et al. (1963) attempted to discriminate between the different kinds of leucoses. They managed to find a slight increase in the fluorescence of peripheral blood lymphocytes in the initial stages of chronic lymphadenosis and in the case of a favorable course of the disease. In cases of pronounced metaplasia in serious forms of the disease, the fluorescence intensity was greatly increased. The authors observed that the main luminescence carrier was the cytoplasm, whereas the nuclei fluoresced weakly. This clearly reflects the topography of the distribution of proteins and nucleic acids in the cell, thus confirming that nucleic acids are not the luminescence center but, on the contrary, an effective screen. Several investigations by Brumberg's school (Brumberg and Barskii, 1960) revealed the opposite effect of malignancy on the luminescence of the nucleus and cytoplasm. In cancer of the uterine cervix the nuclei of the epithelial cells fluoresce more weakly and the cytoplasm more strongly than in the normal case. After the removal of nucleic acids the nuclei of tumor cells (ascites and solid tumors of animals are characterized by brighter fluorescence than normal (Barskii, Zelenin, and Lyapunova, 1962).

The foregoing account shows that investigators have mainly sought a correlation between the luminescence intensity and the pathological process. However, a much greater amount of information about the nature of the structural chemical organization of the protoplasm during the development of the pathological process will be obtained if the comparison involves, not merely the intensity of the fluorescence of various cells, but the magnitude of the changes in fluorescence intensity due to several factors (light, heat, ultrasound, and various substances in different concentrations). I have no doubt that it is fundamentally possible by a choice of an appropriate number of such testing procedures, especially if they include specific reactions of macromolecules which correspond conformationally to one another, such as the antigen-antibody reaction, to compile a very specific and reliable luminescence catalog in which each kind of pathology would be characterized by a specific set of tests with stable indices of change of the luminescence parameters. Of extreme interest in this connection is the work of the Brumbergs (1964), who actually showed a distinct and easily detectable property of the peripheral blood of the healthy organism and the organism at the earliest stages of the carcinogenic process. They found that cells of the transplantable LIO-1 mouse tumor after incubation in the blood plasma of mice with a transplanted tumor have 30% more intense fluorescence than after a similar incubation in the blood plasma of healthy animals. Differences in the ability of the blood of sick and healthy animals to affect the luminescence intensity of tumor cells appeared at early stages in malignant growth. In other .words, at early stages in carcinogenesis the blood plasma undergoes easily detectable changes in its properties, and in it there appears some chemical factor X which is absent in the peripheral blood of healthy individuals.

The clinical significance of such a discovery, if it is subsequently confirmed, can hardly be overestimated. The authors also report that by a similar test they managed to detect differences in the blood of healthy humans and cancer patients. The ability of the blood to quench the fluorescence of cells of Ehrlich's mouse ascites carcinoma returned after successful surgical treatment.

A similar method was used in the work of the Brumbergs (1964), who managed to find distinct differences in the reduction of the fluorescence intensity of segment-nuclear leucocytes and lymphocytes under the action of glycolytic poisons (monoiodoacetic acid and sodium fluoride). This test was successfully used to demonstrate the existence of, and to differentiate between, two kinds of acute leucosis, lymphoidal and myeloidal.

What has been said in this chapter provides some hope that the use of a more varied assortment of luminescence parameters, particularly a quantitative study of the responses of the cell luminescence to a successfully selected set of experimental treatments, will prove in the near future to be a very useful auxiliary instrument in the field of molecular and cell biology and primarily in the field of practical medicine.

Chapter 6

ELECTRONIC EXCITED STATES OF BIOPOLYMERS AND PHOTOBIOLOGY

In this chapter we shall not dwell in detail on the role of electronic excited states of proteins and nucleic acids in photobiological processes (it has been very convincingly shown by numerous investigations) but shall confine ourselves to some remarks of a general nature. An analysis of the published information suggests four main ways in which electronic excited states of biopolymers can be implicated in photobiological cell reactions.

1. Ultraviolet quantum → S*[1] thymine of DNA → T*[2] thymine of DNA → dimerization → defective synthesis of RNA and protein.

2. Ultraviolet quantum → S* protein (aromatic amino acids) → photochemistry → interaction of photochemical products with DNA.

3. Ultraviolet quantum → S* protein (aromatic amino acids) → conformational changes → protein photochemistry → damage to cell function.

4. Ultraviolet quantum → S* protein (aromatic amino acids) → migration to chromophore (provitamin D, protochlorophyllide, etc.) → chromophore photochemistry → biological effect.

The role of nucleic acids as an ultraviolet acceptor in the cell in the case of the bactericidal and mutagenic effect of ultraviolet has been shown in many investigations. The role of protein tryptophan as an ultraviolet acceptor in the cell is just as certain. Action spectra with a protein (tryptophan) maximum at 280 mμ have been found, for instance, for the following processes: suppression

[1] S* is the singlet excited state of the molecule.
[2] T* is the triplet excited state of the molecule.

of the power of interference in the influenza virus (Powell and Setlow, 1956), hemagglutination (Tamm and Fluke, 1950), an increase in the latent period of T-1 bacteriophage (Setlow, 1957), an increase in the mutation rate of the fungus *Chaetomium* (McAulay et al., 1945; Ford, 1947; and McAulay and Ford, 1947), photoaxis of the volvox *Platymonas subcordiformis* (Halldal, 1961), inhibited hatching of larvae from nematode eggs (Hollaender et al., 1940), inactivation of spores of *Ustilago zeae* (Landen, 1939), loss of mobility and death of paramecia (Giese and Leighton, 1935), phototropisms of phycomycetes (Delbrück and Shropshire, 1960) and oat coleoptile (Curry et al., 1956), inhibition of formation of spindle and mitotic apparatus in grasshopper neuroblasts (Carlson, 1954), and the formation of a pale spot in the chromosomes in newt cells due to destruction of nucleic acids (Zirkle, 1957; Zirkle et al., 1960; and Brown and Zirkle, 1964).

Other authors have observed the approximate constancy of the quantum yields of biological processes in the region of nucleic acid and protein absorption. This is manifested in the two excitation spectra maxima at 260 and 280 mμ. These relationships have been observed for the suppression of the infectivity of bacteriophages T-1 and T-2 (Zelle and Hollaender, 1955) and the *B. megaterium* virus (Franklin et al., 1953) and the inactivation and induction of *E. coli* K-12 virus, where a small shoulder at 280 mμ appears in addition to the 260-mμ maximum (Franklin, 1954).

Photobiological reactions involving DNA proceed most rapidly through the triplet excited states of this molecule. According to the calculations of Mantione and Pullman (1964), in the triplet state the nature of the distribution of π-electron density facilitates the breakage of the 5-6 double bond of thymine and the dimerization reaction. Beukers and Berends (1961) arrived at the same conclusion on the basis of the reduction of the quantum yield of the photochemical dimerization of thymine under the action of paramagnetic ions, which quench the phosphorescence and, hence, reduce the steady concentration of molecules in the triplet state.

In contrast to nucleic acids, protein photochemistry and, hence, photobiology proceed via singlet excited states of tryptophan. The experiments of Konev and Volotovskii showed that singlet-singlet energy migration in a trypsin tryptophan-fluorescein system, leading to quenching of the protein tryptophan fluorescence, is accompanied by a proportional reduction in the quantum yield of inactivation of the enzyme. In contrast to this, energy transfer from the triplet level of trypsin tryptophan to chrysoidine, leading to selective quenching of protein phosphorescence and the reduction of its τ (the fluorescence is not quenched, i.e., the steady concentration of molecules in the singlet excited state is not altered), is not accompanied by a reduction in the quantum yield of enzyme inactivation. In other words, enzyme photoinactivation does not depend on the concentration of the triplet excited states of tryptophan.

Chapter 7

ELECTRONIC EXCITED STATES OF BIOPOLYMERS AND DARK BIOLOGY

The question of the role of electronic excited states in dark biology can be considered from two points of view.

Firstly, are light quanta involved in normal dark vital activity, and, if so, do they perform any functional biological role, such as physical mechanism of transmission of information within the cell or between cells? Secondly, do any biochemical or physicochemical reactions of molecules in the dark necessitate population of the energy levels which are manifested in spectroscopic investigations? In other words, can any of the stages of activated molecular complexes acting as an intermediate link between the initial and final products of the chemical reaction be regarded as a state of electronic excitation?

We shall discuss briefly the two possibilities.

BIOLUMINESCENCE IN VISIBLE AND ULTRAVIOLET REGIONS

Bioluminescence in the visible region is well known for many systematic groups of organisms. The investigations of Colli and Facchini (1952), Vladimirov and Litvin et al., (1959 to 1960), Tarusov et al. (1960 to 1965), and Konev et al. (1961 to 1965) have shown that light quanta are created in the course of any processes of vital activity. It is true that the intensity of this universal bioluminescence, which is found in microbial, plant, and animal cells, is extremely small, of the order of 10^4 to 10^5 quanta/sec from 1 cm^2 of cells. The luminescence is enhanced in an oxygen atmosphere, is weakened by free-radical inhibitors, and is greatly increased after exposure to ultraviolet or ionizing radiation (Konev, Troitskii, and Katibnikov, 1961, 1964). The bioluminescence is stimulated by very weak doses of ionizing radiation (1 to 10 rad). What mechanism gives rise to this luminescence? The mechanism responsible for the bioluminescence can be provisionally subdivided into two stages: the creation of a large energy quantum by a chemical process, and the excitation by this energy of some energy level of the molecule, which immediately emits a quantum of bioluminescence. We shall not deal with the first phase and shall merely mention that there appears to be no basis for rejecting the presently accepted point of view (Vasil'ev, 1962 and 1964) that the chemical source of the energy of the ultraweak luminescence is recombination of radicals, particularly oxide or hydroperoxide radicals. This fits in well with the fact of reduction of bioluminescence in the presence of a free-radical inhibitor, propyl gallate.

We are more interested in the question of what substance and what level of it is excited by this energy. We postulate that a considerable part of the total ultraweak luminescence is due to the chemiphosphorescence of protein tryptophan. What justification do we have for this? First of all, our knowledge of the photoluminescence of cells indicates that tryptophan residues in protein always win the competition for light quanta introduced from outside. It would be logical to assume that the situation is the same for energy quanta created endogenously during the chemical processes of vital activity.

Secondly, Vasil'ev's systematic investigations (1962 to 1964) have shown that the introduction of foreign molecules capable of efficient photoluminescence into a system where a chemiluminescent reaction occurs leads to sensitized chemiluminescence with the transfer of energy to this longer-wave acceptor. Of all the cell chromophores present in greatest amount (carbohydrates, fats, lipoids, nucleic acids, and proteins), proteins possess the highest fluorescence quantum yield and the lowest energy levels.

Thirdly, the spectral composition of the bioluminescence of the barley root system, as experiments with absorption filters have shown, is very close to that of the chemiluminescence of tryptophan-containing proteins in hydrogen peroxide (Konev, 1964). Hence, in the first and second case the triplet level of tryptophan is excited.

Thus, we postulate that the triplet levels of protein tryptophan are implicated in the production of the universal ultraweak luminescence of cells.

In contrast to the ultraweak visible luminescence, the ultraweak ultraviolet luminescence by the very method of discovery indicates great biological activity. As is known, the mitogenetic rays discovered as long ago as 1923 by Gurvich were detected by their ability to induce cell division. Roughly speaking, when an inductor, such as onion rootlets, was brought close to a detector (yeast suspension in a quartz cuvette), an interaction took place (which could only be mediated by some physical factor) between the inductor and detector cells and stimulated cell division in the latter. The effect did not occur if an ultraviolet-opaque glass was placed between the inductor and detector. Hence, the physical interaction factor was identified as ultraviolet light.

The subsequent fate of mitogenesis was very peculiar. Partially owing to the unusual nature of Gurvich's theoretical ideas and largely owing to the capricious nature of the biological method of detecting ultraviolet quanta, mitogenesis was actually not accepted in the cytological and biophysical literature. Moreover, since the fifties the extensive use of physical detectors of ultraweak luminescence (photomultipliers cooled to low temperatures) led to the detection of bioluminescence only in the visible region of the spectrum, as already mentioned. The ultraviolet region of the spectrum appeared to be "empty." This necessitated a particularly thorough investigation of the detection of bioluminescence in the ultraviolet region, since the biologically active part of the radiation must be concentrated in this region. However, before conducting the technically difficult experiments to detect ultraweak fluxes of ultraviolet bioluminescence, we considered the first thing to do was to simulate mitogenetic radiation and to convince ourselves that cell division could be stimulated by the entry of one or several quanta into the cell. We think that this is the most paradoxical point: In the cell, i.e., a system consisting of 10^{13} to 10^{14} molecules with multiple interduplicated functions, we must accept the possibility that changes affecting this system as a whole can be produced by a single electronic excitation of only one or a few molecules. Hence, we have to postulate the existence of a very efficient amplifying mechanism triggered by one electronic excited state. The possibility of photodimerization of thymine in the only molecules of the cell which do not duplicate one another—the DNA molecule (in the haploid cell in the state of interkinesis there are no two the same in genetic code, i.e., in molecular structure)—must be ruled out, since light fluxes of mitogenic intensity can cause photochemical inactivation of only one cistron of DNA per 10^3 cells.

In our experiments (Konev and Lyskova, 1964) we "created" artificial mitogenetic radiation by attenuating monochromatic light from a mercury lamp with neutral filters (metal grids). The ultimate intensity of the light flux incident on the cells was 10^4 to 10^5 quanta/sec/cm^2 of surface of the cell suspension.

We thought that decisive evidence of the biological activity of ultraweak ultraviolet luminescence could be obtained by the use of a synchronized cell culture. As is known, all the cells of such a culture are in approximately the same functional state. For a certain period after removal of the synchronizing block they are incapable of dividing, and then they simultaneously enter the phase of mitosis. As a result, the number of cells is doubled in a short interval of time. The biological activity of single ultraviolet quanta absorbed by the cell can be assessed from the reduction of the latent period before onset of the first wave of mitosis.

In the experiments we used a liquid culture of the yeast *Torula utilis* synchronized by removal of the nitrogen source (ammonium sulfate) from Rider's medium for 3 hr. The cell density was 3000 to 6000/mm^3. The cell count was made in a Goryaev cell under the microscope.

Irradiated and control suspensions of yeast cells were put into rectangular quartz cuvettes of the SF-4 spectrophotometer. The two cuvettes were placed in a light-tight chamber fitted to the cuvette section of the SF-4 spectrophotometer. The experimental cuvette was irradiated continuously by the 280-mμ mercury line of an SVD-120A lamp with the above-indicated intensity. The optical width of the slit was 0.05 mμ. The light flux was uniformly defocused over the whole surface of the experimental yeast culture.

Figure 58 shows the result of one of the experiments in which a synchronized culture of yeast cells was irradiated with an ultraweak flux of ultraviolet light. There was distinct stimulation of cell division, manifested in a pronounced reduction of the latent period (by a factor of 3.1 on the average). As a result of the first wave, the percentage of divided cells was usually the same in the control and experiment. This indicates that quanta of ultraviolet light can stimulate only cells which are potentially ready for division. The results of 14 similar experiments are given in Table 20.

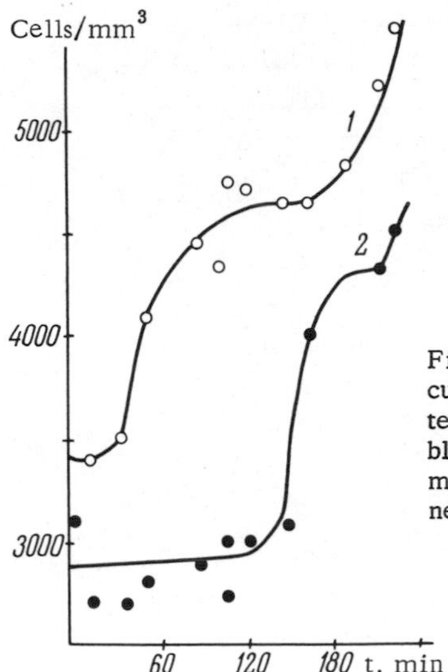

Fig. 58. Change in cell concentration in a synchronized culture of *Torula utilis* in Rider's medium at room temperature in relation to time after removal of the block: (1) illuminated by monochromatic light ($\lambda = 280$ mμ) with dose rate of 10^6 quanta/cm^2/sec; (2) in darkness (Konev and Lyskova, 1965A).

The absence of overlap of the series of latent periods in the experiment and control shows that the observed difference is perfectly significant and makes statistical analysis superfluous.

In our experimental setup a suspension containing about 10^7 cells absorbed $10^6 \times 30 \times 60 = 2 \times 10^8$ quanta in half an hour, i.e., each cell absorbed approximately ten quanta during the whole exposure.

Hence, these experiments demonstrate the fundamental possibility of physical interaction between cells separated by relatively large distances by means of ultraviolet radiation of mitogenetic intensity.

Another two facts are of interest. Firstly, an increase in the dose rate of 10^9 quanta/cm^2/sec led to the disappearance of the stimulating effect. This had been pointed out several times before by Gurvich's school. Secondly, when the irradiated volume of the cells was gradually reduced (by diaphragming of a parallel light beam) a critical-mass effect was observed: Light fluxes covering an area of more than 1 mm^2 were equally effective, whereas light fluxes covering an area of less than 1 mm^2 (10^5 cells) were ineffective. This indicates that the division signal is transferred from cell to cell by a chain mechanism. When the critical mass is attained, the system reacts according to the "all or nothing" rule, even when less than 0.01 of the number of cells is irradiated.

Table 20. Changes in Latent Period (Minutes) of the First Wave of Division of Synchronized Culture of *Torula utilis* under Action of Ultraweak Ultraviolet Radiation (10^6 quanta/cm^2/sec, λ = 280 mμ) (Konev and Lyskova, 1965A)

Latent period, min		
Culture in darkness	Culture in same conditions + ultraviolet	Latent period reduced by factor of
135	25.0	5.40
120	50.0	2.40
120	30.0	4.00
100	60.0	1.70
135	30.0	4.50
135	40.0	3.40
150	35.0	4.30
130	60.0	2.15
180	45.0	4.00
100	30.0	3.30
100	40.0	2.50
120	40.0	4.00
80	20.0	4.00
120	50.0	2.40
Arithmetic mean 123	39.5	3.10

We can surmise that the mechanism of secondary induction consists of a chain process of initiation and transmission from cell to cell of secondary ultraviolet bioluminescence, which causes the stimulation of mitosis.

These experiments confirmed the practicality of attempts to detect mitogenetic radiation of cells by means of a physical detector.

The ultraviolet bioluminescence in the initial experiments (Troitskii, Konev, and Katibnikov, 1961) was detected by means of a self-quenching gas photon counter (Geiger counter) with a platinum photocathode sensitive only to the spectral region of 200 to 300 mμ (Shelkov, Prager, and Kostin, 1959). The quartz window of the counter had an area of about 1 cm^2. The background was reduced by a lead-aluminum shield and an anticoincidence circuit with a ring of shielding counters. The shielding ring consisted of 12 STS-6 tubes arranged in the form of a dome over the photon counter. The operation of the apparatus was made more reliable by increasing the length of the pulse from the shielding counters to several microseconds in the fourth channel of the "Yablonya" apparatus, and the pulses from the photon counter in the first to second channels were retarded 1 μsec by a constant delay line and shortened to fractions of a microsecond. This reduced the background from 50 to 60 to 5 to 8 pulses/min. With this apparatus we were able to detect the ultraviolet bioluminescence of several objects (Table 21).

In addition, we managed to detect the luminescence of the beating frog heart (35% increase over background), which disappeared when the heart stopped, and the luminescence of minced frog muscle (20% over background). No luminescence could be detected in undamaged muscle in a state of physiological rest. If the object, which was placed close (1 to 5 mm) to the quartz window, was replaced by an empty cuvette or a cuvette containing nutrient medium or water, the counting rate was not higher than the background rate. Objects screened with black paper or thin glass also showed a counting rate typical of the background. Thus,

Table 21. Relative Intensity of Ultraviolet Bioluminescence of Some Biological Objects (Counts per Minute) (Troitskii, Konev, and Katibnikov, 1961)

Object	Arithmetic mean ± standard error		
	Preparation	Background	Increase over background
Day-old culture of alcohol yeasts (race 12) in wort	12 ± 1.7	8.3 ± 0.7	47 ± 22
The same on wort agar	7.2 ± 0.1	5.2 ± 0.5	42 ± 10
Dandelion inflorescence in sunny weather	6.44 ± 0.11	5.35 ± 0.18	20.2 ± 3.9
The same in dull weather	5.4 ± 0.18	5.35 ± 0.18	None
The same after ultraviolet irradiation	6.3 ± 0.1	5.35 ± 0.18	17.8 ± 3.6

the luminescence of a whole series of classical mitogenetic objects can be detected with a physical detector. However, the mere establishment of the fact of extremely weak ultraviolet bioluminescence still does not demonstrate the functional, informational role of the quanta produced by the cell.

It would be quite feasible to postulate that these quanta, created by recombination of radicals, merely accompany vital processes and are physiologically unnecessary waste products of the chemistry of life. Hence, we had to find out first of all if cell division is actually accompanied by the emission of ultraviolet quanta. To discover the interrelationship between the ultraweak ultraviolet bioluminescence and cell division, we conducted experiments with a synchronized yeast culture. Synchronization increases the steady concentration of cells simultaneously in the division phase by more than two orders, on the one hand, and, on the other, allows the detection of the bioluminescence of a large assembly of cells in the same stages of the mitotic cycle.

The experiments were conducted in the following way (Konev, Lyskova, and Nisenbaum, 1966): A culture of *Torula utilis* grown in Reader medium was synchronized by removal of ammonium sulfate from the nutrient medium for 3 hr. The addition of the salt was followed by a 2-hr latent period, after which the first wave of budding began. Immediately after removal of the block the yeast culture in a concentration of 10,000 to 30,000 cells/mm^3 was put into a quartz cuvette and placed close to the photocathode of an FÉU-18A photomultiplier cooled through the mount with liquid nitrogen. To determine the sensitivity of the apparatus, we used ZhS-19 luminescent glass, which contains small admixtures of a radioactive substance (uranium). The photomultiplier was operated as a photon counter. The signal was amplified by a D-4 broad-band pulse amplifier and delivered to a PS-10,000 scaler. The average background of the apparatus was 4 to 8 counts/10 sec, and the sensitivity, according to the ZhS-19 standard, was 200 to 300 counts/10 sec. The luminescence of yeasts in darkness in front of the photocathode was measured alternately with the background or with a starved culture of the same yeasts as a control. The averaged result of ten experiments, which all gave similar results, is shown in Fig. 59. This figure shows that in the first few minutes after removal of the block the culture hardly radiated. About 20 min after removal of the block, appreciable emission began and reached a maximum by the fiftieth

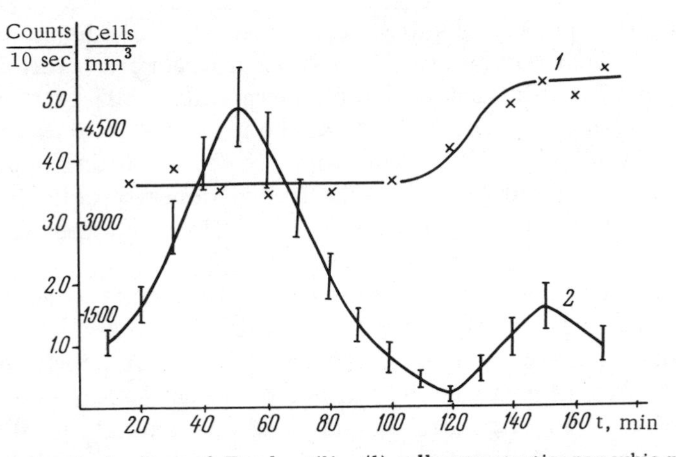

Fig. 59. Synchronized culture of *Torula utilis:* (1) cell concentration per cubic millimeter of suspension in relation to time after removal of block; (2) intensity of ultraviolet bioluminescence of culture (absolute increase over background in counts per 10 sec for background of 6 counts/10 sec) in relation to time after removal of block (Konev, Lyskova, and Nisenbaum, 1966).

minute. The increase over the background at the maximum was 20 to 150% (average, 70%) in different experiments.

The statistical significance of the increase in counting rate over that of the background or of a starved culture was very high, $t = 3.3$, which is significant at the 0.01 level.

The luminescence subsequently declined gradually, and after 2 hr the increase over the background had almost disappeared. After 2.5 to 3 hr there was a second wave of emission, corresponding to the second wave of budding. A comparison of plots of the luminescence intensity against time after removal of the block with the curves of cell number against time showed that the wave of radiation preceded the budding wave, determined from the number of adult cells, by almost 1 hr. Hence, radiation is produced in yeast cells during the time of preparation and development of the division processes at the molecular level, before the appearance of the formed morphological elements. These experiments reveal a very distinct causal relationship between bioluminescence and cell division. Cells in the state of interkinesis (starved cells) lack any detectable luminescence. The fact that radiation actually lies in the ultraviolet region of the spectrum is shown by experiments with absorption filters. For instance, a

BS-8 filter, which transmits light with wavelengths in excess of 380 mμ, almost completely (70%) stops mitogenetic radiation.

A UFS-2 filter (250 to 380 mμ) transmits 80% of the emitted quanta, and a BS-5 filter (330 to 800 mμ), about 50%. Hence, the radiation consists of a broad band between 250 and 380 mμ with a maximum at about 330 mμ. The fluorescence of cells is reduced by a factor of 1.7 to 1.8 by a BS-5 filter and lies in the same region.

During the first wave of mitosis about 100 counts were recorded. If the quantum sensitivity of the photocathode is taken as 100 quanta per count, the solid angle as 40°, and the factor by which the cell luminescence is screened by the proteins and nucleic acids as 0.1, the total number of generated quanta is approximately 10^8. On the assumption that 10^7 cells divide during this time in this system, then each cell emits about 10 quanta during preparation for mitosis. However, on the basis of the obtained data, it is difficult to say if their emission is simultaneous or extends over a more or less prolonged period.

Thus, these experiments can be regarded as a direct confirmation of the existence of mitogenic radiation by physical techniques.

It is of fundamental importance to determine what molecules in electronic excited states in the cell initiate the chain of processes leading ultimately to the biological macroeffect. An investigation of the spectral dependence of the reduction of the latent period in a synchronized cell culture can help to answer this question. For a single-cell layer, Lyskova and I found that light with wavelength 280 mμ, which is absorbed mainly by proteins, has a much lower threshold intensity than light of wavelength 254, 265, 289, or 296 mμ. This indicates that proteins, and not nucleic acids, are the acceptors of the ultraweak bioluminescence of cells.

This conclusion is in good agreement with the observations of Carlson et al. (1961), who treated cells of grasshopper-embryo neuroblasts with ultraviolet microbeams. Using monochromatic microbeams 3 μ in diameter, the authors observed inhibition of mitosis by large doses of ultraviolet. In the case of small doses, however, which cause hardly any significant damage at the molecular level, the exact opposite biological effect was observed: Mitosis was stimulated instead of being inhibited. By determining the interval of time between early to middle prophase, on the one hand, and metaphase, on the other, for "twin" cells of the same embryo in identical stages of mitosis, the authors found that out of 12 cells exposed to weak illumination with 280 mμ in early prophase, 10 attained metaphase earlier than in the control and only 2 later than the control; out of 8 cells which received weak illumination in middle prophase, 7 attained metaphase earlier than the control and 1 at the same time as the control. This stimulating effect of weak illumination on the rate of mitosis was not observed when monochromatic light of wavelength 265 mμ was used.

Thus, protein excitation most probably has a stimulating effect on mitosis.

This gives rise to the suggestion that the stimulating effect of small doses of ionizing radiation on vital activity is due to the induction of ultraweak ultraviolet radioluminescence and the corresponding excited states of proteins.

Finally, a few words on the most obscure point in the mechanism of action of single ultraviolet quanta on the cell, the amplification of the effect. As a working hypothesis we can assume that in the cell as a whole there may occur slight cooperative structural changes in the majority (or all) of the protein molecules initiated by some photochemically (or photophysicochemically) deformed single protein macromolecule. The chain of events leading to cell division can be represented as follows: Light quantum + tryptophan residue of any protein molecule → change in conformation of macromolecule → cooperative change in conformation of neighboring molecules → change in enzyme activity ··· → cell division. The conformational change is apparently not the helix-random coil transition, but is of a finer structural nature. Thus, the present hypothesis assumes that the large assembly of cell proteins can exist as a system in several conformational states, which are almost equivalent energetically to one another, but are separated by some energy barrier. It is also postulated that external factors, including light quanta, can convert cell biopolymers from one conformation to another. The existence of such a structural "phasic" cooperative transition of the cell as a whole would also throw light on the action of some chemical substances (colicins), one molecule of which, though not interacting with any of the forms of nucleic acid, is nevertheless capable of stopping protein synthesis simultaneously in thousands of cell ribosomes.

It is possible that similar mechanisms operate, for instance, in the case of the action of some hormones on the cell. In addition to

Jacob and Monod's well-known scheme, which represents a very specific, but relatively slow biochemical mechanism of autoregulation, the postulated transconformational transitions of proteins in the cell as an integral system can be regarded as an additional, not very specific, but rapid mechanism of intra- and intercellular regulation.

POSSIBLE IMPLICATION OF OPTICAL ENERGY TRANSITIONS IN CHEMICAL REACTIONS

The coincidence of the spectra of photo-, radio-, electro-, and chemiluminescence of organic molecules is of deep significance and indicates that on interacting with different forms of energy the organic molecule uses the same energy levels, which depend on its internal structure. Hence, the molecules must use the optical energy levels to effect movements of their electrons and nuclei, which ultimately determine the electronic nature of the chemical reaction. In this connection the deep internal unity between photochemical and chemical reactions probably extends much further than we can imagine at present. In all probability not only unilateral, but bilateral interrelationships should also be sought. Particular chemical transformations may occur not only as a result of the introduction of energy from outside in the form of a light quantum, but in the course of chemical reactions themselves the exothermic stages may select electronic excited states of molecules as the most efficient pathway for the particular reaction. The occurrence of such a reaction need not necessarily be accompanied by luminescence: The rate constant of the chemical pathway of deactivation of such molecules may greatly exceed the rate constant of the optical radiative pathway of deactivation. There is no need, either, to overcome the whole energy "distance" between the levels at the expense of some single concentrated energy source, such as recombination of radicals, decomposition of hydrogen peroxide, etc. The common glowworm refutes this with its chemiluminescence, where the energy of high-energy phosphorus bonds, not exceeding 12 kcal, gives rise to excited states, corresponding to an energy of 50 to 60 kcal. In fact, one of the initial products of the last phase of the reactions may contain a large supply of free energy, and conditions may be created in the cell for the "assembly" of a large energy quantum from successive small ones with very high efficiency (the quantum yield of conversion of the chemical energy of ATP into light energy is about 90%). One of the greatest successes in this direction was the discovery by Vladimirov and L'vova (1964) of the luminescence accompanying oxidative phosphorylation in mitochondria. These authors showed that, during this universal dark process of extracting energy from food, electronic excited molecules with a free energy supply of 60 to 70 kcal are produced naturally.

We can surmise (following Reid, 1960) that the specific pathway of chemical transformations and the nature of the "sympathies" and the "antipathies" of the molecules to particular types of chemical transformations, as in the case of photochemical reactions, is largely determined by the energy level of the molecule, the structure of its energy levels, the spatial anisotropy, and the probabilities of electronic transitions. Perturbations of the electron density can be compared with a steering wheel, the rotation of which determines the direction of motion of a heavy bus, the atomic nuclei contained in the molecule.

The energy distance between the ground state of substance A and the electronic level of the metastable intermediate state of the chemical reaction determines in this case the minimum energy required for the occurrence of the reaction, the activation energy. The direction of the dipole moment of the transition between the energy states of molecule A and the metastable intermediate product will determine the "trajectory" of the movements of the electron from the molecule in the case of complete or partial ionization—excitation. From this point of view, each molecule has selective allowed directions of displacement of the electrons relative to the skeleton of the molecule, which, together with the vectorial relations of the supplied energy (e.g., the direction of collision), determine the probability of occurrence of the reaction, the steric factor in the kinetics of chemical reactions. Recent years have brought evidence of this. We shall confine ourselves to a few examples.

It has gradually become obvious that processes accompanied by the transfer of electrons from one molecule to the other are by no means limited to classical oxidation-reduction systems but play a much wider and more universal role and determine the ability of molecules and macromolecules to interact in a wide range of biochemical processes.

The role of tryptophan as an electron donor is postulated in many charge-transfer complexes of direct biological significance. Of great interest from this viewpoint is the possibility of specific combination of enzymes with coenzymes, which possess a pyrimidine ring, by the mechanism of a charge-transfer complex. This is indicated, for instance, by the appearance of a broad absorption band when glyceraldehyde-3-phosphodehydrogenase is mixed with nicotinamide dinucleotide (Kosower, 1956; and Cilento and Ginsti, 1959). A similar complex is formed when nicotinamide benzochloride is added to indole or its derivatives. This is manifested in the instantaneous appearance of a yellow color, similar to that observed in an enzyme-coenzyme complex (Cilento and Tedeschi, 1961). The appearance of a broad absorption band with a maximum at 325 mμ, belonging to the charge-transfer complex N-(β-indolylethyl)-3-carboxamide pyridinium chloride, as Shifrin (1964) showed, is accompanied by the hypochromic effect in the region of absorption of indole itself. In our view, the most interesting point in Shifrin's results is that this reduction of absorption is greatest in the spectral regions where the ^1L_a oscillator makes the greatest relative contribution. In the region of absorption of the ^1L_b oscillator there is appreciable reduction of the hypochromic effect, so that at the maximum at 287 mμ the difference spectrum has zero values. These results suggest that only the electrons which are implicated in the transition to the ^1L_a level can move from indole to the pyrimidine ring on the formation of the A^+B^- complex. The orientation of this biochemically active oscillator relative to the skeleton of the molecule and the corresponding absorption spectrum agrees well with the same characteristics of the usual optical oscillator of indole. Hence, we can put forward the highly hypothetical view, which is important for dark biology, that the main biological energy process, biological oxidation, is effected with the participation of the optical oscillators of tryptophan: The oxidized coenzyme is connected with the tryptophan of the enzyme by the mechanism of charge transport in a charge-transfer complex. The optical oscillators of tryptophan probably play a role in a much larger number of biological processes than we can imagine at present. One cannot fail to be amazed by the fact that even before Mulliken, who established the existence and gave the theory of charge-transfer complexes (1950 and 1952), the possibility of the combination of biological molecules with one another, not by chemical bonds, but by the formation of donor-acceptor complexes was advanced by the biochemist Brackmann (1949) to explain the main immunological reaction, the antigen-antibody reaction. This point of view was recently supported by Weinbach (1964). Similar types of reactions, according to Kanner and Kozloff (1964), are the reaction of T-2 phage with indole and of T-4 phage with tryptophan, which makes the phage incapable of penetrating *E. coli* cells. The authors show that combination with the phage tail does not require any functional groups of the indole, imidazole, benzimidazole, or other kinds of rings, by means of which chemical combination might take place. This reaction depends directly only on the electron-donor properties of the inhibitor.

In this case the inhibitor emerges as a factor which, by the formation of a charge-transfer complex involving one of its oscillators, causes a structural change in the protein tail of phages, which renders them incapable of attaching themselves to cells of the host microbe.

Thus, the optical electronic oscillators of the indole ring of tryptophan may play a significant role in the natural biochemical functioning of this molecule by determining the nature of the implication of the indole ring in such fundamental reactions as the enzymatic process and immunological and mechanochemical reactions.

REFERENCES

Agroskin, L.S., and I. Ya. Barskii, Dokl. Akad. Nauk SSSR, 139:987, 1961.
Agroskin, L.S., N.V. Korolev, et al., Dokl. Akad. Nauk SSSR, 131:1440, 1960.
Agroskin, L.S., N.V. Korolev, et al., Dokl. Akad. Nauk SSSR, 136:226, 1961.
Agroskin, L.S., and N.V. Korolev, Biofizika, 6:4, 1961.
Agroskin, L.S., and N.A. Pomoshchnikova, Biofizika, 7:292, 1962.
Augenstein, L.G., and J. Nag-Chaudhuri, Nature, 203(4950):1145, 1964.
Augenstein, L.G., J.G. Carter, D.R. Nelson, and H.P. Jockey, Radiation Res. Suppl., 2:19, 1960.
Azumi, T., and S.P. McClynn, J. Chem. Phys., 37:2413, 1962.
Baba, H., and S. Suzuki, Bull. Chem. Soc. Japan, 35:683, 1962.
Bakhshiev, N.G., Izv. Akad. Nauk SSSR, Ser. Fiz., 24:587, 1960.
Barenboim, G.M., Biofizika, 7:227, 1962.
Barenboim, G.M., Biofizika, 8:321, 1963.
Barenboim, G.M., and A. N. Domanskii, Biofizika, 8:321, 1963.
Barenboim, G.M., I. Ya. Barskii, and R. I. Pinto, Radiobiologiya, 1:843, 1961.
Barenboim, G.M., A.N. Domanskii, and K.I. Pravdina, Proc. Fifteenth Conf. Spectr., Minsk, 1963.
Barenboim, G.M., R.I. Pinto, and K.I. Pravdina, Radiobiologiya, 3:8. 1963.
Barskii, I. Ya., Biokhimiya, 24:823, 1959.
Barskii, I. Ya., E.M. Brumberg, V.K. Vasilevskaya, and G.F. Pluzhnikova, Botan. Zh., 44:639, 1959.
Barskii, I. Ya., E.M, Brumberg, and T.M. Kondrat'eva, Biofizika, 6:605, 1961.
Barskii, I. Ya., E.M. Brumberg, and V.A. Brumberg, Dokl. Akad. Nauk SSSR, 147:474, 1962.
Barskii, I. Ya., A.V. Zelenin, and E.A. Lyapunova, Abstr. Proc. Eleventh Conf. Luminescence, Minsk, 1962.
Beaven, G.H., Advan. Spectr., 2:332, 1961.
Becker, P.I., and P. Szendrö, Pflügers Arch., 228:755, 1931.
Bell, L.N., and G.L. Merinova, Biofizika, 6:159, 1961.
Benesch, R.E., and R. Benesch, J. Am. Chem. Soc., 77:5877, 1955.
Bersohn, R., and I. Isenberg, Biochim. Biophys. Res. Commun., 13:205, 1963.
Bersohn, R., and I. Isenberg, J. Chem. Phys., 40:3175, 1964.
Beukers, R., and W. Berends, Biochim. Biophys. Acta, 49:181, 1961.
Blumenfeld, O.O., and M. Levy, Arch. Biochem. Biophys., 76(1):97, 1958.
Blyumenfel'd, L.A., Izv. Akad. Nauk SSSR, Ser. Fiz., 3:287, 1957.
Blyumenfel'd, L.A., and A.É. Kalmanson, Primary and Initial Processes in the Biological Action of Radiation, Moscow, 1963.
Boaz, H., and G.K. Rollefson, J. Am. Chem. Soc., 72:3435, 1950.
Bobrovich, V.P., and S.V. Konev, Dokl. Akad. Nauk SSSR, 155:197, 1964.
Bobrovich, V.P., and S.V. Konev, Abstr. Proc. Thirteenth Conf. Luminescence, Moscow, 1964A.
Bobrovich, V.P., and S.V. Konev, Dokl. Akad. Nauk BSSR, 9:118, 1965.
Boroff, D.A., Intern. Arch. Allergy Appl. Immunol., 15:74, 1959.

Boroff, D.A., and J.E. Fitzgerald, Nature, 181:751, 1958.
Boroff, D.A., and B.R. Das Gupta, J. Biol. Chem., 239:3694, 1964.
Bowman, R.L., P.A. Caufield, and S. Udenfriend, Science, 122(2157):32, 1955.
Brackmann, W., Rec. Trav. Chim., 68:147, 1949.
Brady, R.O., Biochim. Biophys. Acta, 70:467, 1963.
Braines, S.N., S.V. Konev, G.P. Golubeva, E.V. Kuchina, and O. Ya. Kobrinskaya, in: Questions of Experimental Pathology, Moscow, 1959.
Brand, L., J. Everse, and N.O. Kaplan, Biochemistry, 1:423, 1962.
Brealey, G.J., and M. Kasha, J. Am. Chem. Soc., 77:4462, 1955.
Bredereck, H., G. Höchele, and W. Huber, Chem. Ber., 86:1971, 1953.
Bresler, S.E., Introduction to Molecular Biology, Moscow, 1963.
Brown, D.Q., and R.E. Zirkle, Fourth Intern. Photobiol. Congress, Oxford, 1964, p. 137.
Brumberg, E.M., Zh. Obshch. Biol., 17:6, 1956.
Brumberg, E.M., and I. Ya. Barskii, Tsitologiya, 2:318, 1960.
Brumberg, E.M., and I. Ya. Barskii, Zh. Obshch. Biol., 22:459, 1961.
Brumberg, E.M., and I.E. Brumberg, Biofizika, 9:748, 1964.
Brumberg, E.M., M.N. Meisel, I. Ya. Barskii, and M.P. Bukhman, Zh. Obshch. Biol. 19:99, 1958.
Brumberg, E.M., I. Ya. Barskii, and M.S. Shudel', Tsitologiya, 2:589, 1960.
Brumberg, E.M., I. Ya. Barskii, T.M. Kondrat'eva, and N.A. Chernogryadskaya, Biofizika, 6:114, 1961.
Brumberg, E.M., I. Ya. Barskii, T.M. Kondrat'eva, N.A. Chernogryadskaya, and M.S. Shudel', Dokl. Akad. Nauk SSSR, 135:1521, 1961.
Brumberg, E.M., M.N. Meisel', I.Ya. Barskii, N.V. Zelenin, and E.A. Lyapunova, Dokl. Akad. Nauk SSSR, 141:723, 1961.
Brumberg, E.M., I.Ya. Barskii, N.A. Chernogryadskaya, and M.S. Shudel', Dokl. Akad. Nauk SSSR, 150:1356, 1963.
Brumberg, E.M., I. Ya. Barskii, N.A. Chernogryadskaya, and M.S. Shudel', Izv. Akad. Nauk SSSR, Ser. Biol., 1:87, 1963A.
Brumberg, I.E., Biofizika, 9:382, 1964.
Brumberg, I.E., and E.M. Brumberg, Biofizika, 9:502, 1964.
Burshtein, É.A., Abstr. Proc. Sci. Conf. Young Scientists Biol. and Soil Faculty of MGU, 1960.
Burshtein, É.A., Biofizika, 6:753, 1961.
Burshtein, É.A., Dissertation: The Use of Spectral Methods in the Investigation of the Functional States of Proteins, Moscow, 1964.
Burshtein, É.A., and T.B. Suslova, Biofizika, 9:48, 1964.
Carlson, J.G., M.E. Gaulden, and J. Jagger, Progress in Photobiology, Proc. Third Intern. Congr., Copenhagen, 1960, American Elsevier, New York, 1961, p. 251.
Cherkasov, A.S., Izv. Akad. Nauk SSSR, Ser. Fiz., 24:591, 1960.
Cherkasov, A.S., Opt. i Spektroskopiya, 12:73, 1962.
Chernitskii, E.A., and S.V. Konev, Dokl. Akad. Nauk BSSR, 8:258, 1964.
Chernitskii, E.A., and S.V. Konev, Zh. Prikl. Spektroskopii, 2:261, 1965.
Chernitskii, E.A., S.V. Konev, and V.P. Bobrovich, Dokl. Akad. Nauk BSSR, 7:628, 1963.
Chernogryadskaya, N.A., and M.S. Shudel', Dokl. Akad. Nauk SSSR, 145:917, 1962.
Chizhikova, Z.A., The Radioluminescence Yield of Organic Substances, Consultants Bureau, New York, 1962. (Russian edition: Tr. Fiz. Inst. Akad. Nauk, 15:178, 1961.)
Cilento, G., and P. Giusti, J. Am. Chem. Soc., 81:3801, 1959.
Cilento, G., and P. Tedeschi, J. Biol. Chem., 236:907, 1961.
Cochran, D.R., A. Buzzell, and M.A. Lauffer, Biochim. Biophys. Acta, 55:755, 1962.
Coggeshall, N.D., and G.M. Lang, J. Am. Chem. Soc., 70:3283, 1948.
Colli, L., and V. Facchini, Nuovo Cimento, 11:255, 1954.
Colli, L., V. Facchini, G. Guidotti, R.D. Lonati, M. Orsenigo, and O. Sommarivo, Experientia, 11:479, 1955.
Cornog, J.L., and W.R. Adams, Jr., Biochim. Biophys. Acta, 66:356, 1963.
Cowgill, R.W., Biochim. Biophys. Acta, 75:272, 1963.

Cowgill, R.W., Arch. Biochem. Biophys., 100:36, 1963A.
Cowgill, R.W., Arch. Biochem. Biophys., 104:84, 1964.
Cowgill, R.W., Biochem. Biophys. Res. Commun., 16:322, 1964A.
Cowgill, R.W., Biochim. Biophys. Acta, 94:74, 1965.
Cowgill, R.W., Biochim. Biophys. Acta, 94:81, 1965.
Crammer, I.L., and A. Neuberger, Biochem. J., 37:302, 1943.
Curry, G.M., K.V. Thimann, and P.M. Ray, Physiol. Plantarum, 9:429, 1956.
Debye, P., and I.O. Edwards, Science, 116:143, 1952.
Debye, P., and I.O. Edwards, J. Chim. Phys., 20:236, 1952.
Delbrück, M., and W. Shropshire, Plant Physiol., 35:194, 1960.
Derkosch, J., O.E. Polansky, E. Rieger, and G. Derflanger, Monatsch. Chemie, 92b:1132, 1961.
Dhéré, C., in: Hanbuch der biologischen Arbeitsmethoden, 2nd ed., Vol. 2, No. 4, 1933, p. 3097.
Dhéré, C., La fluorescence en biochemie, Paris, 1937.
Dodd, R.E., and G.W. Stephenson, in: Hydrogen Bonding, Pergamon Press, New York, 1959.
Donovan, I.W., J. Laskowski, and H.A. Scheraga, J. Am. Chem. Soc., 82:2154, 1960.
Douzou, P., and J.M. Thuillier, J. Chim. Phys., 57:96, 1960.
Douzou, P., J.C. Francq, M. Hanss, and M. Ptak, J. Chim. Phys., 58:926, 1961.
Duggan, D., and S. Udenfriend, J. Biol. Chem., 223:313, 1956.
Duggan, D., R. Bowmann, B.B. Brodie, and S. Udenfriend, Arch. Biochem. Biophys., 68:1, 1957.
Edelhoch, H., and R.F. Steiner, Biochim. Biophys. Acta, 60:365, 1962.
Edelhoch, H., and R.F. Steiner, in: Electronic Aspect of Biochemistry, New York, 1964.
Edelhoch, H., L. Brand, and M. Wilchek, Israel J. Chem., 1:216, 1963.
Edsall, J.T., G. Felsenfeld, de Witt S. Goodman, and F.R.N. Gurd, J. Am. Chem. Soc., 76:3054, 1954.
Éidus, L. Kh., in: Physicochemical and Structural Bases of Biological Phenomena, Moscow, 1960.
Éidus, L. Kh., and L.P. Kayushin, Dokl. Akad. Nauk SSSR, 135:1525, 1960.
Eigen, M., and G.G. Hammes, Advan. in Enzymol., 25:1, 1963.
Eisenberg, M., and G. Schwert, J. Gen. Physiol., 34:583, 1951.
Ermolaev, V.L., and I.P. Kotlyar, Opt. i Spektroskopiya, 9:351, 1960.
Euler, H., K. Brandt, and C. Neumüller, Biochem. Z., 281:206, 1935.
Fasman, G., K. Norland, and A. Pesce, Biopolymers Symp. No. 1, p. 325, 1964.
Feitelson, J., J. Phys. Chem., 68:391, 1964.
Feofilov, P.P., The Physical Basis of Polarized Emission, Polarized Luminescence of Atoms, Molecules, and Crystals, Consultants Bureau, New York, 1961. (Russian edition, 1959.)
Feofilov, P.P., and B.Ya. Sveshnikov, Zh. Éksperim. i Teor. Fiz., 10:1372, 1940.
Ford, J.M., J. Gen. Physiol., 30:211, 1947.
Förster, T., Discussions Faraday Soc., 27:7, 1959.
Franklin, R., Biochim. Biophys. Acta, 13:137, 1954.
Franklin, R.M., H. Friedman, and R.B. Setlow, Arch. Biochem. Biophys., 44:289, 1953.
Freed, S., and W. Salmre, Science, 128:1341, 1958.
Freed, S., J.H. Turnbull, and W. Salmre, Nature, 181:1731, 1958.
Fromageot, C., and G. Schenk, Biochim. Biophys. Acta, 6:113, 1950.
Fujimori, E., Biochim. Biophys. Acta, 40:25, 1960.
Gagloev, V.N., and Yu. A. Vladimirov, Abstr. Proc. Conf. Biol. Action Ultraviolet Radiation, Moscow, 1964.
Galanin, M.D., Tr. Fiz. Inst. Akad. Nauk, 5:348, 1950.
Galanin, M.D., Tr. Fiz. Inst. Akad. Nauk, 12:16, 1960.
Gally, J.A., and G.M. Edelman, Biochim. Biophys. Acta, 60:499, 1962.
Gally, J.A., and G.M. Edelman, Biopolymers Symp. No. 1, p. 367, 1964.
Gerstein, J.E., H. Van Vunakis, and L. Levine, Biochemistry, 2:964, 1963.

Giese, A.C., and P.A. Leighton, J. Gen. Physiol. 18:557, 1935.
Giese, A.C., and P.A. Leighton, Science, 85:428, 1957.
Gorin, G.J., Am. Chem. Soc., 78:767, 1956.
Gribova, Z.P., R.P. Evstigneeva, A.F. Mironov, L.P. Kayushin, V.N. Luzgina, and A.K. Piskunov, Biofizika, 8:550, 1963.
Grossweiner, L.I., J. Chem. Phys., 24:1255, 1956.
Grossweiner, L.I., and W.A. Mulac, Radiation Res., 10:515, 1959.
Guroff, G., A. Michael, and M. Chirigos, Anal. Biochemistry, 3:330, 1962.
Gurvich, A.G., and L.D. Gurvich, Introduction to the Science of Mitogenesis, Moscow, 1948.
Gurvich, A.G., and L.D. Gurvich, Die mitognetische Strahlung, Jena, 1959.
Hachimori, I., H. Horinishi, K. Kurihara, and K.Shibata, Biochim. Biophys. Acta, 93:346, 1964.
Halldal, P., Physiol. Plantarum, 14:133, 1961.
Ham, I.S., and J.R. Platt, J. Chem. Phys., 202:335, 1952.
Havsteen, B.H., and G.P. Hess, J. Am. Chem. Soc., 84:491, 1962.
Heckman, R.C., J. Mol. Spectry., 2:27, 1958.
Helmholtz, V., Physiol. Optik, 2:284, 1886.
Hermans, I., Jr., Biochemistry, 2:453, 1963.
Hess, A., Pflügers Arch., 137:339, 1911.
Hess, A., Arch. Regl. Ophthalmol., 23:3, 1911.
Hollaender, A., and M.R. Zelle, in: Progress in Photobiology, Proc. Third Intern. Congr. Photobiol., Copenhagen, 1960, American Elsevier, New York, 1961.
Hollaender, A., M.F. Jones, and L. Jacobs, J. Parasitol., 26:421, 1940.
Hollas, Y.M., Spectrochim. Acta, 19:753, 1963.
Hoshijima, S., Sci. Papers Inst. Phys. Chem. Res. (Tokyo), 20:109, 1933.
Isenberg, I., and R. Bersohn, in: Electronic Aspects of Biochemistry, New York, 1964.
Isenberg, I., and A. Szent-Györgyi, Proc. Natl. Acad. Sci., USA, 44:519, 1958.
Isenberg, I., R.B. Leslie, S.L. Baird, R. Rosenbluth, and R. Bersohn, Proc. Natl. Acad. Sci., USA, 52:379, 1964.
Jones, R.R., Chem. Rev., 41:353, 1947.
Jordan, P., Naturwissenschaften, 26:693, 1938.
Kanner, L.C., and L.M. Kozloff, Biochemistry, 3:215, 1964.
Karreman, G., R. Steele, and A. Szent-Györgyi, Proc. Natl. Acad. Sci., USA, 44:140, 1958.
Kasha, M., Chem. Rev., 41:401, 1947.
Kasha, M., M.A. El-Bayoumi, and W. Rhodes, J. Chim. Phys., 58:916, 1961.
Katibnikov, M.A., and S.V. Konev, Biofizika, 7:150, 1962.
Khan-Magometova, Sh. D., A.V. Gutkina, M.N. Meisel', L.S. Agroskin, and N.V. Korolev, Biofizika, 5:446, 1960.
Khrushchev, N.G., Arkh. Anat., Gistol., i Embriol., 64:39, 1963.
Kohlrausch, K.W.F., and R.Seka, Ber. Deut. Chem. Ges., 71B:1563, 1938.
Konev, S.V., Dissertation: Some Features of Photochemical Transformations in Biological Systems, Moscow, 1957.
Konev, S.V., Dokl. Akad. Nauk SSSR, 116:594, 1957.
Konev, S.V., Byull. Nauchn.-Tekhn. Inform. VIZha, 2:55, 1958.
Konev, S.V., Abstr. Proc. Sixth Conf. Luminescence, Leningrad, 1958A.
Konev, S.V., in: Problems of Photosynthesis, Moscow, 1959.
Konev, S.V., in: Questions of Experimental Pathology, Moscow, 1959.
Konev, S.V., Izv. Akad. Nauk SSSR, Ser. Fiz., 23:89, 1959.
Konev, S.V., Vestsi Akad. Nauk BSSR, 4:59, 1964.
Konev, S.V., Vestn., Ser. Khim. Nauki, 9:144, 1964.
Konev, S.V., Abstr. Proc. Conf. Biol. Action Ultraviolet Radiation, Vilna, 1964.
Konev, S.V., Abstr. Proc. First Biochem. Congr. No. 2, Moscow, 1964.
Konev, S.V., and V.P. Bobrovich, Abstr. Proc. Conf. Biol. Action Ultraviolet Radiation, Vilna, 1964.
Konev, S.V., and V.P. Bobrovich, Biofizika, 10:813, 1965.

Konev, S.V., and E.A. Chernitskii, Biofizika, 9:520, 1964.
Konev, S.V., and E.A. Chernitskii, Abstr. Proc. Conf. Biol. Action Ultraviolet Radiation, Vilna, 1964A.
Konev, S.V., and M.A. Katibnikov, Dokl. Akad. Nauk SSSR, 136:472, 1961.
Konev, S.V., and M.A. Katibnikov, Biofizika, 6:638, 1961.
Konev, S.V., and M.A. Katibnikov, Abstr. Proc. Symp. Bioluminescence, Moscow, 1963.
Konev, S.V., and T.I. Lyskova, Biofizika, 8:260, 1963.
Konev, S.V., and P.N. Saloshenko, Dokl. Akad. Nauk BSSR, 7:696, 1963.
Konev, S.V., N.A. Troitskii, and M.A. Katibnikov, Sectl. Repts. Fifth Intern. Biochem. Congr., 2:35, 1961.
Konev, S.V., M.A. Katibnikov, and T.I. Lyskova, Abstr. Proc. Eleventh Conf. Luminescence, Moscow, 1962.
Konev, S.V., T.I. Lyskova, and V.P. Bobrovich, Biofizika, 8:433, 1963.
Konev, S.V., M.A. Katibnikov, and V.P. Bobrovich, Abstr. Proc. Thirteenth Conf. Luminescence, Moscow, 1964.
Konev, S.V., E.A. Chernitskii, V.P. Bobrovich, and M.A. Katibnikov, Abstr. Proc. Thirteenth Conf. Luminescence, Moscow, 1964.
Konev, S.V., M.A. Katibnikov, and T.I. Lyskova, Biofizika, 9:124, 1964.
Konev, S.V., T.I. Lyskova, and P.N. Saloshenko, Dokl. Akad. Nauk BSSR, 8:51, 1964.
Konev, S.V., N.A. Troitskii, and M.A. Katibnikov, Vestsi Akad. Nauk BSSR, Ser. Biyal., 1:76, 1964.
Konev, S.V., V.P. Bobrovich, and E.A. Chernitskii, Biofizika, 10:42, 1965.
Konev, S.V., V.P. Bobrovich, and E.A. Chernitskii, Dokl. Akad. Nauk SSSR, 165:937, 1965A.
Konev, S.V., V.P. Bobrovich, and T.I. Lyskova, in: Cell Biophysics, Moscow, 1965, p. 98.
Konev, S.V., E.A. Chernitskii, and V.R. Muraveinik, Dokl. Akad. Nauk SSSR, 10:341, 1965.
Koshland, D.E., Jr., in: The Enzymes (P.D. Boyer, ed.), Academic Press, New York, 1959, Vol. 1, Chap. 7.
Koshland, D.E., J.A. Jankelov, Jr., and J.A. Thomas, Federation Proc., 21:1031, 1962.
Kosower, E.M., J. Am. Chem. Soc., 78:3497, 1956.
Kulin, E.T., Dokl. Akad. Nauk SSSR, 4:78, 1960.
Kulin, E.T., in: Bioluminescence, Moscow, 1965, pp. 21 and 196.
Laskowski, M., S.I. Leach, and H.A. Scheraga, J. Am. Chem. Soc., 82:571, 1960.
Lehrer, S.S., and G.D. Fasman, Biopolymers, 2:199, 1964.
Leighton, W.G., and P.A. Leighton, Can. J. Chem., 12:139, 1935.
Levshin, V.L., Photoluminescence of Liquids and Solids, Moscow, 1951.
Levshin, V.L., and L.V. Krotova, Opt. i Spektroskopiya, 13:809, 1962.
Lewis, G.N., and M. Kasha, Can. J. Chem., 66:2100, 1944.
Linschitz, H., M. G. Berry, and D. Schweitzer, J. Am. Chem. Soc., 76:5833, 1954.
Lippert, E., in: Luminescence of Organic and Inorganic Materials (H.P. Kallmann and G.M. Spruch, eds.), Interscience, New York, 1962.
Lippert, E., W. Luder, and H. Boos, in: Advances in Molecular Spectroscopy (A. Mangini, ed.), Pergamon Press, New York, 1962, p. 443.
Longin, P., Compt. Rend., 248:1971, 1959.
Longworth, J.N., Biochem. J., 81:23, 1961.
Longworth, J.N., Biochem. J., 84:104, 1962.
Loofbourow, I., Cook, and Stimson, Nature, 142:573, 1938.
Lumry, R., S. Yanari, and F.A. Bovey, Fourth Intern. Photobiol. Congr., Oxford, 1964.
Magill, M.A., R.E. Steiger, and A.I. Allen, J. Biochem., 31:188, 1937.
Mantione, M.J., and B. Pullman, Biochim. Biophys. Acta, 91:387, 1964.
McAulay, A.L., and J.M. Ford, Heredity, I:247, 1947.
McAulay, A.L., N.J.B. Plaulay, and J.M. Ford, Australian J. Exp. Biol. Med. Sci., 23:53, 1945.
McLaren, A.D., and D. Shugar, Photochemistry of Proteins and Nucleic Acids, Pergamon Press, New York, 1964.
Mizushima, S., M. Tsuboi, T. Shimonouchi, and Y. Tsuda, Spectrochim. Acta, 7:100, 1955.
Mulliken, R.S., J. Am. Chem. Soc., 72:600, 1950.

Mulliken, R.S., J. Chem. Phys., 19:514, 1951.
Mulliken, R.S., J. Am. Chem. Soc., 74:811, 1952.
Nagakura, S., and H. J. Baba, J. Am. Chem. Soc., 74:811, 1952.
Nagakura, S., and H. Gouterman, J. Chem. Phys., 26:881, 1957.
Nag-Chaudhuri, J., and L. Augenstein, Biopolymers Symp. No. 1, p. 441, 1964.
Neporent, B.S., and N.G. Bakhshiev, Opt. i Spektroskopiya, 8:777, 1960.
Neyroth, H., and I. Loofbourow, J. Am. Chem. Soc., 53:3441, 1931.
Omel'chenko, S.I., Z.V. Pushkareva, and S.G. Bogomolov, Zh. Obshch. Khim., 27:3220, 1957.
Pariser, R., J. Chem. Phys., 24:250, 1956.
Patten, F., and W. Gordy, Proc. Natl. Acad. Sci., USA, 46:1137, 1960.
Perlmann, G.E., Biopolymers Symp. No. 1, p. 383, 1964.
Pesce, A., E. Bodenheimer, K. Norland, and G.D. Fasman, J. Am. Chem. Soc., 86:5669, 1964.
Pikulik, L.G., M. Ya. Kostko, S.V. Konev, and E.A. Chernitskii, Dokl. Akad. Nauk BSSR, 10:6, 1966.
Pil'shchik, E.M., and M.V. Nikolaeva, Dokl. Akad. Nauk SSSR, 148:199, 1963.
Pimentel, C.C., J. Am. Chem. Soc., 79:3323, 1957.
Platt, J.R., J. Chem. Phys., 17:484, 1949.
Platt, J.R., J. Chem. Phys., 19:101, 1951.
Powell, W.F., and R.B. Setlow, Virology, 2:337, 1956.
Praissman, M., and A. Rupley, J. Am. Chem. Soc., 86:3584, 1964.
Preiss, J.W., and R.B. Setlow, J. Chem. Phys., 25:138, 1956.
Pringsheim, H., and O. Gerngross, Ber. Deutsch. Chem. Ges., 61:2009, 1928.
Ptak, M., and P. Douzou, Nature, 199:1092, 1963.
Pullman, B., and A. Pullman, in: Burton, M., et al., Comparative Effects of Radiation, John Wiley & Sons, New York, 1960.
Raddi, K.K., Biochim. Biophys. Acta, 24:238, 1957.
Rahn, R.O., J.W. Longworth, J. Eisinger, and R.G. Shulman, Proc. Natl. Acad. Sci., USA, 51:1299, 1964.
Reeder, W., and V.E. Nelson, Proc. Soc. Exp. Biol. Med., 45:792, 1940.
Reid, C., Excited States in Chemistry and Biology (Russian transl.), Moscow, 1960.
Rich, A., and M. J. Kasha, J. Am. Chem. Soc., 82:6197, 1960.
Riehl, N., Naturwissenschaften, 43:145, 1956.
Roe, E.M.F., Photoelec. Spectrometry Group Bull., 13:371, 1961.
Rosenheck, K., and G. Weber, Biochem. J., 79:29, 1961.
Rozanova, L.M., I. Ya. Barskii, and E.M. Brumberg, Dokl. Akad. Nauk SSSR, 150:907, 1963.
Saidel, L.J., Arch. Biochem. Biophys., 54:184, 1955.
Sannigrahi, A.B., and A.K. Chandra, J. Chem. Phys., 67:1106, 1963.
Sapezhinskii, I.I., and N.M. Émanuel, Dokl. Akad. Nauk SSSR, 131:1168, 1960.
Sapezhinskii, I.I., and N.M. Émanuel, in: Bioluminescence, Moscow, 1965.
Sapezhinskii, I.I., and Yu. V. Silaev, in: Bioluminescence, Moscow, 1965.
Sapezhinskii, I.I., Yu. V. Silaev, and N.M. Émanuel, in: Bioluminescence, Moscow, 1965.
Schantz, E.J., D. Stefanye, and L. Spero, J. Biol. Chem., 235:3489, 1960.
Scheraga, H.A., and I.A. Rupley, Advan. Enzymol., 24:161, 1962.
Schur, J., Montsh. Chem., 88:1017, 1957.
Schütt, H.U., and H. Zimmermann, Ber. Bunsen Ges. Phys. Chemie, 67:54, 1963.
Setlow, R.B., Advan. Biol. Medical Phys., 5:37, 1957.
Setlow, R.B., and W.L. Carrier, Fourth Intern. Photobiol. Congr., Oxford, 1964.
Sevchenko, A.N., Dokl. Akad. Nauk SSSR, 42:349, 1944.
Sevchenko, A.N., S.V. Konev, and M.A. Katibnikov, Dokl. Akad. Nauk SSSR, 153:875, 1963.
Shelkov, L.S., I.A. Prager, and A.G. Kostin, Pribory i Tekhn. Éksperim., 3:57, 1959.
Shifrin, S., Biochim. Biophys. Acta, 81:205, 1964.
Shore, V.G., and A.B. Pardee, Arch. Biochem. Biophys., 60:100, 1956.

Shpol'skii, É. V., Uspekhi Fiz. Nauk, 77:321, 1962.
Shtrankfel'd, I. G., Biofizika, 8:690, 1963.
Shtrankfel'd, I. G., Dokl. Akad. Nauk SSSR, 154:953, 1964.
Shtrankfel'd, I. G., Dokl. Akad. Nauk SSSR, 155:461, 1964A.
Shudel', M. S., N. A. Chernogryadskaya, V. A. Brumberg, Yu. A. Rozanov, and E. M. Brumberg, Dokl. Akad. Nauk SSSR, 157:1904, 1964.
Smaller, B., Nature, 195:593, 1962.
Stadler, L. J., and F. M. Uber, Genetics, 27:84, 1942.
Stauff, G., Angew. Chem., 3:151, 1964.
Steele, R. H., and A. Szent-Györgyi, Proc. Natl. Acad. Sci., USA, 43:477, 1957.
Steele, R. H., and A. Szent-Györgyi, Proc. Natl. Acad. Sci., USA, 44:540, 1958.
Steiner, R. F., and H. Edelhoch, Nature, 192:873, 1961.
Steiner, R. F., and H. Edelhoch, Nature, 193:87, 1962.
Steiner, R. F., and H. Edelhoch, Nature, 193:375, 1962A.
Steiner, R. F., and H. Edelhoch, J. Biol. Chem., 238:925, 1963.
Steiner, R. F., and H. Edelhoch, Biochim. Biophys. Acta, 66:341, 1963A.
Steiner, R. F., R. E. Lippoldt, H. Edelhoch, and V. Frattali, Biopolymers Symp. No. 1, p. 355, 1964.
Stimson, M., and M. Reater, J. Am. Chem. Soc., 63:697, 1941.
Stracher, A., Federation Proc., 19:1, 1960.
Stracher, A., J. Biol. Chem., 235:2302, 1960A.
Stryer, L., Biochim. Biophys. Acta, 35:242, 1949.
Stryer, L., Radiation Res. Suppl., 2:432, 1960.
Stübel, P., Arch. Ges. Physiol., 1:142, 1911.
Sturtevant, J. M., Biochem. Biophys. Res. Commun., 8:321, 1962.
Suzuki, S., and H. Baba, J. Chem. Phys., 38:349, 1963.
Sveshnikov, B. Ya., Dissertation: Phosphorescence of Organic Compounds, GOI, Leningrad, 1951.
Szent-Györgyi, A., Science, 93:609, 1941.
Szent-Györgyi, A., Biochim. Biophys. Acta, 16:167, 1955.
Tamm, I., and D. J. Fluke, J. Bacteriol., 59:449, 1950.
Tanford, C., and M. Wagner, J. Am. Chem. Soc., 76:333, 1954.
Tarusov, B. N., A. I. Polivoda, and A. I. Zhuravlev, Radiobiologiya, 1:150, 1961.
Tarusov, B. N., A. I. Polivoda, A. I. Zhuravlev, and E. N. Sekalova, Tsitologiya, 4:696, 1962.
Teale, F. W. J., Biochem. J., 76:18, 1960.
Teale, F. W. J., Photoelec. Spectrometry Group Bull., 13:346, 1961.
Teale, F. W. J., and G. Weber, Biochem. J., 65:476, 1957.
Teale, F. W. J., and G. Weber, Biochem. J., 72:369, 1958.
Terenin, A. N., Photochemistry of Dyes and Related Organic Compounds, Moscow, 1947.
Terenin, A. N., and V. L. Ermolaev, Dokl. Akad. Nauk SSSR, 85:547, 1952.
Thorne, C. J. R., and N. O. Kaplan, J. Biol. Chem., 238:1861, 1963.
Tramer, Z., and D. Shugar, Acta Biochim. Polon., 6:235, 1959.
Troitskii, N. A., S. V. Konev, and M. A. Katibnikov, Biofizika, 6:238, 1961.
Udenfriend, S., Fluorescence Assay in Biology and Medicine, Academic Press, New York, 1962.
Ungar, G., and D. V. Romano, Proc. Soc. Exp. Biol. Med., 97:324, 1958.
Ungar, G., and D. V. Romano, J. Gen. Physiol., 46:267, 1962.
Ungar, G., E. Aschheim, S. Psychoyos, and D. V. Romano, J. Gen. Physiol., 40:635, 1957.
Van Waegeningh, J. E. H., and J. E. Heesterman, Chem. Weekblad., 29:134, 138, 650, 1932.
Vasil'ev, R. F., Dokl. Akad. Nauk SSSR, 144:143, 1962.
Vasil'ev, R. F., Dissertation: Luminescence in Chemical Reactions in Solutions, Moscow, 1964.
Vasil'ev, R. F., in: Bioluminescence, Moscow, 1965.
Vasil'ev, R. F., and A. A. Vichutinskii, Izv. Akad. Nauk SSSR, Ser. Fiz., 2:729, 1963.
Velick, S. F., J. Biol. Chem., 233:1455, 1958.

Velick, S.F., in: A Symposium on Light and Life (W.D. McElroy and Bentley Glass, eds.), Johns Hopkins University Press, Baltimore, 1961.
Velick, S.F., C.W. Parker, and H.N. Eisen, Proc. Natl. Acad. Sci., USA, 43:477, 1957.
Vladimirov, Yu. A., Dokl. Akad. Nauk SSSR, 116:780, 1957.
Vladimirov, Yu. A., Izv. Akad. Nauk SSSR, Ser. Fiz., 23:86, 1959.
Vladimirov, Yu. A., Dokl. Akad. Nauk SSSR, 136:960, 1961.
Vladimirov, Yu. A., Fourth Intern. Photobiol. Congr., Oxford, 1964.
Vladimirov, Yu. A., and É.A. Burshtein, Biofizika, 5:385, 1960.
Vladimirov, Yu. A., and S.V. Konev, Biofizika, 2:3, 1957.
Vladimirov, Yu. A., and S.V. Konev, Biofizika, 4:533, 1959.
Vladimirov, Yu. A., and Li Chin-kuo, Biofizika, 7:270, 1962.
Vladimirov, Yu. A., and F.F. Litvin, Biofizika, 4:601, 1959.
Vladimirov, Yu. A., and F.F. Litvin, Biofizika, 5:127, 1960.
Vladimirov, Yu. A., and D.I. Roshchupkin, Biofizika, 9:282, 1964.
Vladimirov, Yu. A., F.F. Litvin, and Tan Man-ch'i, Biofizika, 7:675, 1962.
Wada, A., and Y. Ueno, Biopolymers Symp. No. 1, p. 343, 1964.
Wang, Shu-fang, F.S. Kavahura, and P. Talalay, J. Biol. Chem., 238:576, 1963.
Weber, G., Advances in Protein Chemistry, Vol. 8, Academic Press, New York, 1953.
Weber, G., Trans. Faraday Soc., 30:552, 1954.
Weber, G., Biochem. J., 75:335, 1960.
Weber, G., Biochem. J., 75:345, 1960.
Weber, G., Nature, 470:27, 1961.
Weber, G., in: A Symposium on Light and Life (W.D. McElroy and Bentley Glass, eds.), Johns Hopkins University Press, Baltimore, 1961.
Weber, G., Appl. Spectry., 17:131, 1963.
Weber, G., and K. Rosenheck, Biopolymers Symp. No. 1, p. 333, 1964.
Weber, G., and F.W.J. Teale, Biochem. J., 72:369, 1958.
Weber, G., and F.W.J. Teale, Discussions Faraday Soc., 27:134, 1959.
Weil, G., and M. Calvin, Biopolymers, 1:401, 1963.
Weinbach, R., Nature, 202:409, 1964.
Weinberg, C.I., D.R. Nelson, J.G. Carter, and L.G. Augenstein, J. Chem. Phys., 36:2869, 1962.
Wels, P., Pflügers Arch., 219:738, 1928.
Wels, P., Pflügers Arch., 227:562, 1931.
Wels, P., Pflügers Arch., 228:671, 1931.
Wetlaufer, D.B., Advan. Protein Chem., 17:303, 1962.
White, A., Biochem. J., 71:217, 1959.
White, A., Doctoral Thesis, Univ. Sheffield, 1960.
Williams, R., J. Chem. Phys., 30:233, 1959.
Yagi, K., and T. Ozawa, Biochim. Biophys. Acta, 62:397, 1962.
Yang, J., and J. Foster, J. Am. Chem. Soc., 77:2374, 1955.
Zelle, M.R., and A. Hollaender, Radiation Biology, Vol. 2, New York, 1955.
Zelle, M.R., and A. Hollaender, J. Bacteriol., 68:210, 1964.
Zimmermann, H., and H. Geisenfelder, Z. Elektrochem., 65:368, 1961.
Zimmermann, H., and N. Joop, Z. Elektrochem., 64:1215, 1960.
Zimmermann, H., and N. Joop, Z. Elektrochem., 65:61, 1961.
Zirkle, R.E., Advan. Biol. Med. Phys., 5:103, 1957.
Zirkle, R.E., and R.B. Uretz, Ann. N. Y. Acad. Sci., 90:435, 1960.

INDEX